# PROTEIN

# THE BASIS OF ALL LIFE

*By the same Author*

THE BIOLOGY AND TREATMENT OF VENEREAL DISEASE
(Harrison & Sons, 1915)

LINKS IN A CHAIN OF RESEARCH ON SYPHILIS
Hunterial Lectures (Harrison & Sons, 1916)

VENEREAL DISEASES
(Heinemann, 1920)

THE NATURE OF DISEASE
(Heinemann, 6 Vols., 1924-34)

THE COMMON COLD AND INFLUENZA
(Heinemann, 1936)

THE UNIVERSE THROUGH MEDICINE
(Heinemann, 1940)

THE NATURE OF DISEASE UP TO DATE
(Heinemann, 1947)

THE NATURE OF DISEASE INSTITUTE'S FIRST ANNUAL REPORT
(Heinemann, 1948)

THE NATURE OF DISEASE INSTITUTE'S SECOND ANNUAL REPORT
(Heinemann, 1950)

THE NATURE OF DISEASE INSTITUTE'S THIRD ANNUAL REPORT
(Heinemann, 1951)

A FURTHER STUDY IN THE NATURE OF DISEASE
(Heinemann, 1954)

A FINAL STUDY IN THE NATURE OF DISEASE
(Heinemann, 1959)

# PROTEIN

## *The Basis of All Life*

J. E. R. McDONAGH, F.R.C.S.

WILLIAM HEINEMANN MEDICAL BOOKS LTD
LONDON

*First published* 1966



Printed in Great Britain by
The Whitefriars Press Ltd., London and Tonbridge

# CONTENTS

# FOREWORD

This book, the final record of a man who had dedicated his life to the study of health and disease, was completed only a few weeks before his death. It represents his attempt, after his retirement, to summarise the main findings and conclusions which over 50 years of clinical and laboratory work had prompted him to reach.

It is for me a special privilege to write this foreword. As a boy I had no vocational attraction to medical work and ironically, since I was an only son, I was to prove the first break in eight generations of McDonaghs to follow a medical career. For the time being at least, therefore, there is no-one in a position to pursue and develop my father's ideas. This is the more serious because, in contrast to the current trends towards specialisation and differentiation, the results of his work convinced him that for a correct understanding of the nature of the universe, health and disease, a *correlative* approach is required.

The more experience my father gained, the more confirmed he became of the soundness of his unitary theory of medicine. "It is wrong," he used to say to me "to refer to diseases. You really mean different manifestations of disease. There is only one basic disease, that is when the colloid protein particles in the blood of man or in the sap of plants become over-expanded or over-contracted."

Now to follow an unorthodox path, certainly in professional spheres, is rarely rewarding and often builds up opposition. My father's experience was no exception to this. It is, therefore, my earnest hope and plea that scientists will take up this book with an open mind. The author himself would realise that perseverance will be required before at least parts of the book can be digested. But perhaps future readers will come to realise that whilst my father's views may not be, probably are not, wholly right, they are assuredly not wholly wrong.

The fundamental conclusions reached by the author can be summarised as follows:

1. That every product of which the universe is composed is made up of three portions; the store, the radiant and the attractive. They are linked by junctions or bridges.

2. That the behaviour of every product can be reduced to the three functions of storing, radiating and attracting.

3. That this cyclical behaviour pattern shows itself by "activity," in the course of which the portions alternately expand and contract, or pulsate; and the products themselves progress and regress.

4. That the progressive products include the atomic, chemical, colloid, vegetable and animal, the third group acting as a bridge between the two inanimate and the two animate groups. Examples of the regressive products are the sub-atomic particles; the fungi, bacilli and cocci; and the spore-bearing *Protozoa*.

5. That health and disease are synonyms in the animate products for progress and for regress in the inanimate products.

6. That none of these four states of progress and regress, health and disease, is divisible from the others. For descriptive purposes disease may be divided into three stages (the acute, subacute and chronic); into two morbid conditions (inflammation and cancer); and into innumerable ever-changing manifestations depending on the degree of over-expansion and/or over-contraction, which the colloid protein particles in the blood or sap are made to undergo.

7. That the prevention of disease is secured by keeping the colloid protein particles in balance; and, that the treatment of disease induced by this balance being disturbed should aim at restoring to normal the abnormal physico-chemical change undergone.

8. That this "activity" responsible for maintaining and disturbing the protein balance is radiated mainly from the atoms in the hottest stars, slightly from those in the sun, and additionally from those now being split by man.

J. P. McDONAGH

*August* 1966

# PREFACE

BEING no longer able to continue my daily routine of examining patients by the bedside; their blood, along with the bacterial content of their excreta and that of the soil; and of attempting to unravel the chemistry of the glands, belonging to the integrated endocrine glandular system in the laboratory, I set to work to condense the views a fifty-three years long combination of clinical and laboratory work had made me hold on the " Nature of the Universe, Health and Disease."

This title, that came to be borne by the resulting paperback, derived from the combination making me form the opinion, that every product of which the universe is composed is connected with every other product; and that the connexion arises through every product having been formed as a result of energy, or " activity " condensing. Every primary, or whole product, is alike in being composed of three portions; the store, the radiant and the attractive. The portions receive their names from the functions they exhibit, the store the function of storing, the radiant the function of radiation and the attractive the function of attraction. The function of storing is unreactive compared with those of radiation and attraction, both of which are reactive. The store is the primary portion, as it is the one from which the radiant portion developed. The attractive portion developed from the radiant portion, and it was in order to keep balanced the reactive function of radiation that the attractive portion came to exhibit the opposite reactive function of attraction.

Both the development of the portions and the exhibition of their functions followed the description by energy, or "activity," describing cycles in the products of its condensation, which having once begun have ever since continued to be described. Whence comes this energy, or " activity," has yet to be ascertained for certain; but so far as is known at present a little comes from the sun, more from the milky way and most from the outer inter-galactic spaces. The description of the cycle is characterised by the store and attractive portions alternately expanding and contracting; and the radiant portion contracting and expanding. These physico-chemical changes are the first pair every product is capable of undergoing; and, in the process of the description of the cycle within it, the product pulsates—hence the name "pulsatory" that has been given to these cycles. The unique features of the description of these pulsatory cycles are; first, that contraction is the first of the two physico-chemical changes the radiant portion undergoes; and, second, that it is through this change being reactive that the attractive portion is rendered reactive on expanding and the store portion on contracting. The contracted store portion is rendered reactive by the transfer of hydrogen atoms from the radiant portion, which occurs on the portion expanding in the second circuit "activity" makes in its description of the pulsatory cycles in the products.

The uniqueness of these features is demonstrated in the development of secondary cycles from the primary, or pulsatory, cycles, in the description of which the radiant portion is seen to take the lead of the other two portions. But despite the lead gained by the radiant portion, the results are tempered by the expansion and contraction the store portion is made to undergo in the description of the pulsatory cycles, on account of the store portion being made to expand before the radiant portion is able to contract.

The expansion the store portion undergoes in the description of the pulsatory cycles is inconstant. The inconstancy arises from the energy, or "activity," reaching and penetrating the product varying in quality and quantity. One effect of the variation is the division of the pulsatory cycles into diurnal, menstrual, annual and multiennial, in half of which the product expands more than it contracts and in the other half the physico-chemical changes are reversed. For example, in the diurnal cycle the product expands further than it contracts during the day and *vice versa* during the night.

Another effect is the production of the second pair of physico-chemical changes every product is capable of undergoing in the process of the description of the pulsatory cycles within the product. The pair includes the physico-chemical changes of progress and regress. Progress is recorded when more energy is attracted by the attractive portion on expanding and stored in the junction between it and the store portion on its contracting, than is liberated and radiated from and lost to the product on the store and radiant portions expanding, the results of which are regress. The first of this second pair of physico-chemical changes is the one responsible for the condensation, which forms the basis of evolution; and by means of which the components of the universe became connected. The components are divisible into five sections, the atomic, crystalline, colloid, vegetable and animal; the colloid being the one which divides the five sections into the two halves, the inanimate and the animate; in the latter half of which the words progress and regress may be replaced by those of health and disease.

Health like progress is indivisible; the same can be said of regress, but not so of disease. Disease has become divisible through organs having originated from the colloid product protein, which forms the basis of the animate products, and through air and food being required to keep in harmony the relationship that thereby became established between the parent and its offspring. It was the establishment of this relationship that caused the development of the secondary, or metabolic, cycles from the primary, or pulsatory cycles, and that revealed in a way that was previously impossible, the lead the radiant portion of protein takes of the store and attractive portions in the description of the cycles.

No sooner had "The Nature of the Universe, Health and Disease" been published, than I realised the tremendous gap that had been allowed to form in the crystalline section. The moment I realised that this gap had arisen from my having failed to keep abreast with the advances that had been made in biochemistry, I began to remedy the defect. I had not gone far before the significance of the development of the metabolic from the pulsatory cycles; and of the radiant being the portion of the protein which plays the greatest role in their description, impressed themselves upon me.

The object of this pamphlet is to fill the gap, and to show where the attempt fits into the views expressed in this Preface, the majority of which have resulted from the daily examinations that had been made of the protein in the blood over a period of many years. The tests considered to throw the greatest light on the physical state in which the colloid protein particles in the serum stand, were the refractive index, viscosity, so-called "percentage" of the protein and the indices, V.R.I. (viscosity-refractive index) and V.P.I. (viscosity-percentage of protein index). A fall in the first three tests was held to be indicative of the extent to which the protein particles had been made to expand, and in the indices the extent to which this physical change was relative to the contraction the particles had been made to undergo; and a rise to the contraction and relative contraction the particles had been made to undergo.

The charting of the results produced, combined with the clinical examinations of the patients from whom the blood had been obtained, were the factors responsible for the division of disease into stages, morbid conditions and manifestations.

The stages were found to be three in number, the acute, in which the protein particles are relatively over-expanded, the subacute, in which they are relatively over-contracted and the chronic, in which the particles are over-expanded and relatively over-expanded. The test which proved most useful in bringing out the difference between the acute and chronic stages, was the determination of the rate at which the red blood corpuscles fall in a column of citrated plasma—it being found to be increased in the acute stage and unaltered in the chronic. The difference was held to be due to the reflection of the contraction the protein particles undergo in the acute stage, in addition to the expansion, on to the red blood which, after initially expanding, proved to contract, in the process of which they form clumps, or go into *rouleaux* formation.

The morbid conditions number two, inflammation and cancer; and refer to the physical changes the junction between the store and radiant portions, and the organs which originated from it, undergo in the three stages; inflammation in the acute and subacute and cancer in the chronic.

The main physical change the junction undergoes in the acute stage is over-contraction, the result of which is the formation of what have come to be called viruses. The corresponding change to occur in the subacute stage is over-expansion. In this change so-called " positively charged " atoms and groups are released from the protein particles which have been sent into solution; and following their release they proceed to restore the particles to the colloid state, in the process of which the particles clump together, or go into *rouleaux* formation and clot.

In the chronic stage it is an organ which originated from the junction, rather than the region itself, which is affected physically. The organ is the nucleolus of the nucleus of the cells in the organ attacked, which is made first to over-contract and later alternately to over-expand and over-contract as it multiplies and forms metastases. The metastases being regressive forms of the product from which they originated, are apt to develop parasitic qualities.

The organ in which the manifestation appears is governed by the principal abnormal physical change the protein particles are made to undergo. The more relatively over-expanded are the particles in the acute stage, the more likely is the respiratory system to be the one attacked; and the less relatively over-expanded, the nervous system; the portal and cutaneous systems falling in between, in the order mentioned.

The relative over-contraction the particles undergo in the subacute stage, accounts for the manifestation making its appearance either in the tissues or vascular system or both, and to be of a rheumatic nature, as the organs concerned were those which originated from the store portion.

The organ to be attacked in cancer is governed by the relative over-expansion and over-contraction the store portion has been made to undergo in the acute and subacute stages, as these physical changes are those which are primarily responsible for the over-expansion and relative over-expansion the store portion undergoes in the chronic stage.

The agent playing the greatest role in the production of the relative over-contraction the store portion undergoes in the subacute stage, is the micro-organism resident in the large intestine, which has been rendered pathogenic by the relative over-expansion the portion has been made to undergo in the acute stage. The main cause of this relative over-expansion is the same "activity," or "climate," that is responsible for the description of the pulsatory cycles in the particles. Confirming the last statement, is the finding of viruses in the potential victims of an acute, or seasonal, manifestation of disease; and another, the probability of the over-contraction the nucleolus may be made to undergo in cancer, resulting in the production of a virus.

The conclusions reached on the manifestations, suggested the aim of treatment being to correct the abnormal physico-chemical changes their causes had made the protein particles undergo. As the action of the remedies is perhaps more chemical than physical; and, chemistry rather than physics is the subject-matter forming the basis of this pamphlet, the principles of the treatment of the manifestations of disease are dealt with later.

From the light this combination of clinical and laboratory work appeared to throw on the nature of protein, no difficulty arose in surmising, in the case of the atomic products, that the metals are those which exhibit predominantly the function of radiation, the non-metals the one of attraction and the inert gases the function of storing; and that radio-activity is the result of the over-contraction which the store portion is made to undergo causing the radiant portion to over-expand.

The questions arising demanding answers were: Why the metallic atoms so far outnumber the non-metallic and inert gases. And why, considering the radiant portion developed from the store portion, no evidence is to be found of the latter in the most important of all the atoms, hydrogen? The answers I was prompted to give, were that the higher is climbed the elemental ladder the more relatively over-contracted the store portion of the atoms tends to become. And that in the hydrogen atom no sharp line of demarcation can be drawn in the junction between where the store portion, from which the neutron is extracted, ends and the radiant portion, from which the proton is obtained, begins.

I saw a likeness between the subatomic products and the viruses, the latter of which are built up of varying sized ends of the store portion and beginnings of the radiant portion, which made me suggest that the products are regressive forms of the atoms rather than the bricks which are used in their construction. On the principles being enunciated no difficulty was experienced in deciding, that the crystalline products came to be formed by "activity" neutralising the exhibition of the reactive functions of radiation and attraction, and uniting them to exhibit the un-reactive function of storing. Which, having been the first of the functions to be exhibited, has apparently continued to remain the function most desirous of being exhibited.

Once the function of storing came, in this way, to be exhibited predominantly, the three portions of the crystalline products were enabled to gain an independence, known to be possessed by any earlier evolved product of the condensation of "activity." Advantage was taken of this independence by the store portion developing, aided by the radiant portion, whose independence along with that of the attractive had resulted in the splitting of the water in which the development was taking place, into its constituent parts; i.e. the hydrogen atom and the hydroxyl group. By this atom and group forming two electrical and protective layers around the product, the latter was enabled to evolve in a manner that had previously never been possible. Compounds came to develop into complexes, the wholes and parts of which were able to remain in solution, irrespective of the expansion and contraction they might be made to undergo. The ease with which these physical changes were thus rendered reversible, resulted not only in the crystalline state advancing into the colloid, but also in the independence gained by the three portions in the former becoming replaced by interdependence in the latter.

The first effect of the interdependence gained by the three portions of the emulsion colloids, or emulsoids, was the lead the radiant portion took of the other two; and, one which was shown by atoms and groups being able to leave the radiant portion, to enter the other two portions, and to return to their original portion. The factor responsible for the atoms and groups, constituting the basis of the radiant portion, having a freedom of movement, those constituting the basis of the store and attractive portions do not possess, is the reactiveness of the radiant portion both when it contracts and expands in the description of the pulsatory cycles in the product. Both the store and the attractive portions are reactive only when they expand, on contracting they are rendered unreactive, a state to which the radiant portion is reduced only when it is made to over-contract.

The second effect of the advent of interdependence, was the opportunity it gave for organs to originate from the three portions, an event which resulted in the most advanced of the emulsoid colloids, which was protein, to evolve into the vegetable and animal products.

As might be expected from the lead the radiant portion had gained over the other two portions, it was the portion from which the first organs to evolve originated. The first of these organs to appear was the nucleus; and it is owing to this position it held that the nucleus became unique amongst organs. This uniqueness is shown in two ways; the first, by the nucleus entering the store portion wherein it developed into the female sex, following the nucleus, which had remained in the radiant portion, having developed into the male sex. The second, by the organ becoming a constituent of every cell in every organ, irrespective of the portion from which the organ had originated. Despite the probability that the vegetable and animal products evolved from the same product, each did so in their separate ways. In the case of the vegetable products, the store portion underwent much the same progressive relative over-contraction it had previously done in the atomic section and with much the same result; but instead of subatomic particles being formed, fungi, bacilli, sub-bacilli and cocci appeared. Another, but slower regression the vegetable products underwent, following this abnormal physical change the store portion had been caused to undergo, was the alternation of generations which made its first appearance in the bryophytes and progressed to the seed plants, in the latter of which the embryo sac had come to represent the female gamete and the pollen grains the male gamete.

In the case of the animal products the store portion was better able, through not having been made to undergo an abnormal physical change, to form closer associations with the two reactive portions, with which each of its halves is more closely connected. At some points in the evolution of the animal products, the association the store portion formed with the radiant portion was closer than the one it formed with the attractive portion and the other way round. But that the aim was to preserve a balance is clearly shown by the origin, from the centre of the store portion, of the notochord in the *Chordata*. Despite the development of this organ, the store portion was not prevented from forming a closer association with the radiant than the attractive portion in the birds, and the reverse in the mammals. The former is confirmed by the greater development of the respiratory system, which originated from the radiant portion, and the latter by the greater development of the nervous system, which originated from the attractive portion.

The light these two systems throw on the way protein evolved in the animal products, is not to be compared with that revealed by the sense-organ mechanism, which originated from both the attractive and radiant portions, in the invertebrates.

This mechanism was initiated by the pigmented eyespot, which originated from the upper part of the attractive portion, and came to control the nervous and organ part of the mechanism, the former of which originated from the attractive portion and the latter both from this and the radiant portions.

The pigmented eyespot evolved into the eyes, the nervous part into the autonomic and central nervous systems, and the organic part into the specialised and generalised sense-organs in the invertebrates, and the integrated endocrine glandular system in the vertebrates. The last change was probably more of the nature of regress than progress. But this is not the point, the point being that the specialised sense-organs are the more important of the two, and that the so-called " ductless " glands, into which they developed, originated from the junctions between the three portions of the protein, whilst the generalised sense-organs and the so-called " ductless " glands, into which they developed, originated from the portions themselves. The greater importance of the junctions than the portions, is further emphasised by the calco-phosphatides constituting the basis of the specialised sense-organs and the so-called " ductless " glands, into which they developed, but not of the generalised sense-organs and the so-called " ductless " glands, into which they developed.

# INTRODUCTION

The object of this pamphlet is to show the relationship that exists between the primary, or pulsatory, cycles and the secondary, or metabolic, cycles which " activity " describes in protein.

Owing to the different meanings which have come to be implied in the use of the word " protein," what should be made clear is the sense in which the word is used in the present context. Protein is arbitrarily fixed as having been the most perfect and therefore probably the last product of the condensation of " activity " to have evolved in the colloid section. The product is an emulsion colloid, or emulsoid, and owes its perfection to not undergoing hydration and losing its colloid characteristics as its fellow colloid products are accustomed to do.

The probability of protein having been the last of the products to have evolved in the third, or colloid, section, derives from its further evolution having resulted in its forming the basis of the vegetable and animal products, which evolved in the fourth and fifth sections. In these two last products to evolve, protein formed the basis of the cytoplasm of the unicellular plants and animals, which developed later into the particles in the sap and blood of the multicellular products. Protein is next seen as forming the basis of the three germinal layers of the metazoic animal products. From this last position held by the product, not only may it be learnt that protein is divisible into the same three portions as are the other entire products of the condensation of " activity," but also the portions from which the organs originated, and of which protein formed the final basis.

The three germinal layers include the mesoblast, or mesoderm, which corresponds with the store portion of the protein; the hypoblast, or endoderm, which corresponds with the radiant portion; and the epiblast, or ectoderm, which corresponds with the attractive portion.

The same difficulty arises in connexion with these germinal layers as with the atomic products; namely, to find the mesoblast in the coelenterates, or hollow-gutted animals, and the store portion in the hydrogen atoms. Here again, it is not a matter of there being no mesoblast but only that this layer, along with the hypoblast, are not to be separated in the larval form of these animals as is the case in the hydrogen atom. In the physical sense protein exists either in the form of albumin or globulin, according to whether it be expanded or contracted. The forms of albumin and globulin are almost as multiple as are the degrees of expansion and contraction protein may be made to undergo. The emphasis falls more on globulin and contraction than on albumin and expansion, for the reason, that the chemical compounds constituting the basis of protein are physically more apparent when they are in the colloid than when they are in the crystalline state. On changing from the crystalline into the colloid state, protein assumes a threadlike form, which becomes more globular the further into the colloid state it progresses, or the more contracted it becomes.

The chemical compounds constituting the basis of protein, vary according to the organ from which the product has been taken, and the portion from which the organ originated. But owing to the growing interdependence which is being acquired both by the portions of the protein and the organs which originated from them, the chemical compounds constituting their basis are becoming ever more closely integrated as the product evolves.

The chemical compounds constituting the basis of the store portion of the fully evolved protein product, include sodium chloride, sugars, fats and fat-soluble vitamins and steroids. Those constituting the basis of the radiant portion, potassium chloride, nucleic acids, water-soluble vitamins, porphyrins and amino acids. Those constituting the basis of the attractive portion are the adrenergic and cholinergic compounds; and of the junctions between the portions the calco-phosphatides and -cerebrosides.

The paucity of the compounds constituting the basis of the attractive portion, compared with those constituting the basis of the other two portions, arises from the portion having developed from the radiant portion, and from the part which expands having become independent of the part contracting, which remains part of the radiant portion.

In addition to the portions being connected through the junctions, which act as bridges, they are joined at the centre to which the name " terminus " has been given. So long as the pulsatory and metabolic cycles are capable of being described, the protein may be considered to be alive. The factor responsible for its decease is the terminus over-contracting, on account of this physico-chemical change occurring at the site being the one which renders all three portions unreactive, and any further description of the cycles impossible. The over-contraction is usually preceded by the opposite physico-chemical change of over-expansion, an occurrence which causes the percentage of the urea in the blood to be raised, owing to the area being the one in which the urea cycle is described. The organs which originated from the store portion include the tissues and the vascular and genito-urinary systems. Those which originated from the radiant portion include the nucleus, chloroplasts, blood corpuscles and the respiratory and portal systems. Those which originated from the attractive portion include the skin, eyes and nervous systems.

On account of the integrated endocrine glandular system in the vertebrates having been the one into which the organic part of the sense-organ mechanism in the invertebrates developed, and the specialised part of which originated from the junctions and the generalised from the portions, the organs belonging to the system need to be mentioned separately. Beginning with the first junction, the one between the store and radiant portions, as no members of the system originated from the unreactive portion, the glands which originated from the area are the parathyroids, the sole action of which is that of contraction. The glands originating from the radiant portion include the thymus and the islet cells of Langerhans, both of which have a contractile and expansive action. From the second junction, the one between the radiant and attractive portions, originated the thyroid gland which has solely an expansive action. From the attractive portion itself originated the lower part of the chromaffin system, which has no other than an expansive action. Coming to the third junction, the one between the attractive and store portions, the necessity arises to divide the area into the expansive and contractile parts, from the former of which originated the *Pars intermedia*, the expansive half of the posterior pituitary gland and the anterior pituitary gland, and from the latter of which originated the contractile half of the posterior pituitary gland and the pineal gland and the parietal organ. The *Pars intermedia* and the two halves of the posterior pituitary gland form the upper part of the chromaffin system.

The divisibility of the anterior pituitary gland into the basophilic, neutrophilic and acidophilic parts, arises from the expansive action of the first and third parts being the same as those of the attractive portion and store portions, the difference between which is neutralised by the second part. The expansive action of the store portion is of a disintegrative nature, compared with that of the attractive portion which is one of condensation.

## THE SECONDARY, OR METABOLIC, CYCLES

THESE cycles differ from the primary, or pulsatory, cycles in having to be considered from the chemical instead of from the physical angle, despite the term physico-chemical being the one most generally employed to cover both; in the belief that they are inseparable. Inseparable as are the two kinds of cycles, as the biochemists will discover after making the daily examinations of the blood mentioned above and charting the results obtained; and as I found on attempting to become conversant with the advances the biochemists had made during the time the blood examinations were being undertaken.

Ever since becoming acquainted with these advances, I have done my best to discover where they fit into the physical examinations. In attempting to combine the two I have learnt, in addition, that if light is to be thrown on the way in which treatments acts in helping Nature to restore a sick individual to health, the double, or physico-chemical, approach is the more desirable one to be made.

Looking at the two approaches separately, the physical one shows that the description of the pulsatory cycles is begun by energy, or " activity " entering, penetrating and releasing " activity " from where it is stored in the junction between the attractive and store portions of the protein. On being released the " activity " proceeds to make two circuits in the protein in the first of which the store portion expands, the radiant portion contracts and the attractive portion expands; and in the second of which the three portions undergo the opposite physical changes. On account of the " activity ", which I have come to refer to as " climate " entering the protein in ever varying quality and quantity, the " activity " released from the storage depot, following its penetration, is rendered equally variable; with the result that the three portions of the protein may be made to over-expand and over-contract, instead of being made to undergo their more normal physical changes. So long as the pulsatory cycles are normally described health is maintained; it being only when the three portions are made to over-expand and over-contract that the state becomes replaced by the one of disease. The most significant feature of these primary cycles, is that two circuits need to be made in the product, before a single cycle can be described. The significance lies not only in the observation offering an explanation for light being radiated in quanta, which occurs only when the portion expands; but also for active atoms and groups in the store and attractive portions being accepted by the radiant portion on its contracting and donated to the other two portions on its expanding. This acceptance and donation of active atoms and groups by the radiant portion from and to the store and attractive portions, mark the beginning of the description of the metabolic cycles. In addition, they are the factors responsible for the lead the radiant portion acquires over the other two portions in the description of these cycles. But what is insufficiently taken into account in the search for the source of the energy which sparks off the description, particularly in the vegetable products, is the influence the expansion the store portion undergoes in the description of the pulsatory cycles, has upon the contraction the radiant portion undergoes in that of the metabolic cycles.

As expansion of the store portion is the first physico-chemical change the protein undergoes in the description of both the primary and the secondary cycles; as it results in energy, or "activity" being liberated from the product; and as the active atoms and groups accepted by the radiant portion are those the portion donated to the store portion on expanding, following the contraction the store portion had previously been made to undergo, it follows. First, that the energy, or "activity," needed for the description of the metabolic cycles is rendered available by the earlier description of the pulsatory cycles. And second, despite the lead the radiant portion has over the store and attractive portions in the description of the metabolic cycles, nevertheless, the results produced cannot help being influenced by the manner in which the pulsatory cycles have earlier been described.

Another factor that assisted in giving the radiant portion the lead over the store and attractive portions, other than those that the portion developed from the store portion and the attractive from the radiant portion. And that the radiant portion is reactive on both contracting and expanding, and not solely on undergoing the latter physical change as are the other two portions; was the radiant portion being the one in which the non-metallic atoms nitrogen, phosphorus and sulphur and the metallic atoms magnesium, potassium, calcium, iron and the trace elements were the first to become essential ingredients of the portion. Finally must be mentioned, that the organs most concerned in the description of the metabolic cycles, are those which originated from the radiant portion. Mentioned in the order of their importance, are the chloroplasts and the red blood corpuscles and the respiratory system in the description of what are often referred to as the respiratory cycles; the portal system in the description of the digestive cycles; and the nucleus, from which the sexes developed, in the description of the reproductive cycle. The last mentioned cycle is not as a rule included in the metabolic cycles, but in view of the fact that so many of the chemical compounds involved in the description of the secondary cycles are those constituting the basis of the nucleus, the organ that every cell possesses, its inclusion would appear to be desirable.

Having developed from the pulsatory cycles, in the description of which the products are made either to progress or to regress, the object of the description of the metabolic cycles has automatically come to be to keep this second pair of physico-chemical changes reactive. The keeping reactive has grown to be more complicated, especially as the animate products, in which progress and regress in the inanimate products have come to be replaced by health and disease; and in the maintenance and production of which the radiant portion has played an ever increasingly greater role. There has been no fundamental change; instead of its being a matter of the expansion and contraction the portions undergo, it has grown to be one of the building up and breaking down of the chemical compounds of which the portions are composed. The complications have arisen from protein having been born in water; from plants requiring light and carbon dioxide in order to be able to breathe and chemical products on which to live; and from animals needing oxygen, which the plants have produced, to breathe and vegetable and animal products on which to live.

The initial energy, or " activity " required for the description of the metabolic cycles, continues to be that which is needed for the description of the pulsatory cycles; all the additional energy, or " activity," required has its origin from the synthesis and analysis the chemical compounds undergo in the description of the metabolic cycles. As all the chemical compounds constituting the basis of the three portions of the protein, of the water in which it exists as an emulsion colloid, and of the additional products plants and animals require for respiratory and digestive purposes, are in a perpetual state of being synthesised and degraded. As a close relationship exists between them all; and as the lead possessed by the radiant portion in both halves of the respiratory cycles can be reduced to the first pair of physical changes the portion is made to undergo in the description of the pulsatory cycles, there would appear to be no necessity to refer to the compounds constituting the basis of the radiant portion, in the description of some of the metabolic cycles, as enzymes, and the chemical compounds which activate them as co-enzymes.

The chemical compounds constituting the basis of the radiant portion most responsible for giving the portion the lead it gained over the other two portions, were the amino acids. These chemical compounds developed from the fatty acids in the citric acid cycle, which is described in the store portion. In the radiant portion not only did the amino acids go into sequence to form polypeptide complexes, but also they became the parents of the other chemical compounds which were to complete the basis of the portion. The going into sequence running parallel, as it must have done, with the progress made by the store and attractive portions in the evolution the protein undergoes in the description of the pulsatory cycles, must almost certainly have been the factor responsible for the specificity that came to be established, between the particular sequences into which the amino acids went and the particular metabolisms the chemical compounds, constituting the basis of the other two portions which are governed by them, undergo.

The compounds constituting the basis of the radiant portion, having developed from the amino acids, would naturally be expected to assist the polypeptides in their work of leadership; and, at the same time, to have a group rather than a single specificity, which is what is well known to be the case. In other words, a physico-chemical relationship exists between each complex and the chemical compounds in the store and attractive portions with which it interacts in the description of the metabolic cycles. In the actual description of the cycles little or nothing is known of the manner in which the amino acid complex in the radiant portion, reacts with its substrate in the store and attractive portions; and no sharp division has yet been drawn between the complexes and the catalysts which activate them. These catalysts are either smaller complexes or chemical compounds, which transfer either a hydrogen atom or a group from the radiant portion when it expands to the substrate in the store or attractive portion in its contracted state; and takes them back when the radiant portion contracts and the other two portions expand.

The chemical compounds found most frequently to be concerned in accepting and donating the hydrogen atom are the water-soluble vitamins. And the group found most frequently to find its way into the store portion and to return to the radiant is the phosphate ($H_3PO_4$), belonging to the nucleotides. The hydrogen atom accepted by the water-soluble vitamins may be the one displaced from the substrate following its acceptance of the phosphate group, or one which is donated by the vitamin for the express purpose of stepping up the reactivity of the action. The former role of the hydrogen atom is played more frequently when the substrate is being degraded and the latter role when it is being built up.

The chemical compounds in the store and attractive portions which lose their hydrogen atoms in the process of the portions' expanding, become oxidised, whilst those in the radiant portion which accept these hydrogen atoms in the process of the portion's contracting, become reduced.

In the event of the store portion being made to over-expand, which is the first abnormal physico-chemical change the protein is made to undergo in disease, the radiant portion over-contracts and the attractive portion over-expands. In the over-expansion of the store portion more than the donated hydrogen atom may be loosened from the portion, and taken up by the radiant portion as it over-contracts. This happening could account for the finding of a hydroxy-methyl group in some of the pyrimidines in the viruses, which is not found in the normally contracted nucleotides. Use could be made of this group to bind the compounds constituting the basis of the viruses closer together, and thus render them unreactive. In this unreactive state viruses, like all over-contracted protein, have the inherent urge to be restored to their normal physico-chemical state. The change is most readily effected by the over-contracted protein being brought into contact with the protein responsible for the over-contraction. In the case of the viruses, being brought into contact with normal protein is sufficient to effect the change, which is one of a breaking up of the two proteins rather than of a propagation, or reproduction, of the viruses.

The damage the protein suffers from its attractive portion being over-expanded, is its loss of " activity " which is gained by the product responsible for the abnormal physico-chemical change the portion has been made to undergo. The effect of the over-expansion the radiant portion is made to undergo, following the store portion's earlier over-contraction, is the failure of the donated hydrogen atom and groups to reach their destination. On finding themselves in solution, into which the atom and groups go more readily in the venous than in the arterial circulation (excepting the lungs), they fail to find their way to the portion in the protein to which they should normally go. Wandering loose and being so-called " positively charged," the atom and group play a role in causing the protein particles in the blood to clot. Clotting *in vivo* is the abnormal physico-chemical change the protein particles in the blood undergo in the subacute stage of disease, corresponding to the one of virus and inclusion body formation they, and the cytoplasm of the epithelial and nerve cells, undergo in the acute stage. Whether a virus or an inclusion body is formed, is governed by the relative over-expansion the store portion is made to undergo. The greater the relative over-expansion the more likely is a virus to be formed in the protein in the blood, and the less an inclusion body in the cytoplasm of the epithelial cells, and in the nucleus, in addition, in the nerve cells. In the chronic stage of disease when the store portion is both sufficiently over-expanded and relatively over-expanded as to cause cancer to arise, the resulting over-contraction the radiant portion undergoes is that situated in the nucleolus of the organ attacked.

In congenital disease the store portion suffers more than the radiant portion; over-expansion causing hypertrophic lesions to form in the organs which originated from the portion, and over-contraction, atrophic. In inherited disease the attractive and radiant portions are the two to be more affected than the store portion, and mainly for the reason that the two reactive portions are more closely connected than is either reactive portion with the unreactive portion; thus accounting for the lesions formed being either cutaneous or nervous, or portal.

## THE PHOTOSYNTHETIC CYCLE

This cycle is the most important of all those of a secondary nature, for upon its description depends the synthesis of all the organic chemical compounds constituting the bases of the three portions of protein. The products found to be essential for the description of the cycle are; first, the plant pigments; second, the light radiated by the sun; the third, carbon dioxide; and the fourth, water.

The chief of the plant pigments are the chlorophylls, to be found in the chromoplasts and chloroplasts which originated from the radiant portion; and the fat-soluble vitamins which have their home in the store portion. The two groups of pigments are kept connected by the phytyl tail of the chlorophylls, which made its way into the store portion. The pigments in the chloroplasts are green, whereas those in the chromoplasts to be found in the seaweeds, may be brown, pink, red or purple. The pigments in the store portion may be either orange-yellow or yellow, hence the names "carotenes" and "xanthophylls," which have been given to them. The light energy is accepted by and stored in the pigments, especially chlorophyll a, during the day when the store portion of the protein is more expanded than contracted and the radiant portion is more contracted than expanded. During the night the physical changes these two portions undergo are reversed; and, it is when the radiant portion expands that the stored energy becomes released.

On being released in its chemical form, the energy is both retained in the radiant portion and radiated to the store portion. The retained energy is used for the phosphorylation of adenosine diphosphate and the hydrogenation of nicotinamide-adenine dinucleotide phosphate, two compounds, which are needed to activate carbon dioxide in its synthetic work. The energy radiated to the store portion, presumably by way of the phytyl tail, is accepted by and stored in the fat-soluble vitamins, especially plastoquinone, as the store portion contacts. On this energy being released, which happens when the store portion expands, part, aided by the energy radiated by another chlorophyll, is used for the photolysis of water, and part is returned by means of a cytochrome to chlorophyll a. This light energy is not to be differentiated from the "activity," which is responsible for the description of the pulsatory cycles, in the diurnal one of which the store portion of the protein is more expanded than contracted during the day and the reverse during the night. Owing to plants being fixed to the positions in which they live, compared with animals which are free to alter theirs, the store portion of vegetable protein tends to be relatively more contracted than the corresponding portion in animal protein. In actual fact, it is this greater contraction of the store portion of the vegetable protein that stamps the whole life of plants, and causes it to differ so much from the life of animals.

The compound responsible for fixing carbon dioxide, and enabling it to be converted to carbohydrate, is ribulose diphosphate. The combination of these two compounds, assisted by their two catalysts, nicotinamide-adenine dinucleotide phosphate and adenosine triphosphate, results in the formation of a carbohydrate, which proceeds to split into two compounds of phosphoglyceric acid. Each triose with a phosphate group, obtained from adenosine triphosphate, and catalysed by a pyridine nucleotide in possession of a sulphydryl group, is reduced to its aldehyde form, the form from which the other carbohydrates are synthesised.

Two compounds of triose phosphate combine to form hexose diphosphate which, after losing a phosphate group, may become stored as starch. Or it may interact with triose phosphate to form a tetrose and pentose phosphate, from the latter of which ribulose phosphate is formed. The tetrose phosphate combines with triose phosphate to form heptose diphosphate, which is subsequently dephosphorylated. Finally, a 2-carbon fragment is abstracted from the 2-carbon diphosphate carbohydrate, and made to combine with triose phosphate to yield two compounds of pentose phosphate, from which ribulose diphosphate may later be obtained.

In the photolysis of water, which is an oxidation-reduction process, the compound is plit into the hydrogen atom and the hydroxyl group. The splitting takes place in the store portion, and occurs when the portion expands. The hydrogen attracts " activity " from the hydroxyl group, and is accepted by the pyridine nucleotides in the radiant portion, in which it becomes reduced as the portion contracts. The acceptance by the pyridine causes the nitrogen atom to give up its so-called " positive charge," and the ring to lose its aromatic character. The effect of these changes is to energise the hydrogen, in a way that enables it to act as a catalyst on its being radiated to the store portion when the radiant portion expands. The hydroxyl group remains in the store portion where, following its loss of " activity " to the hydrogen atom, it becomes oxidised. When the store portion expands the group may be split into oxygen and water; and, whilst the expansion lasts the oxygen is liberated to find its way into the air. But when the portion contracts use is made of the oxygen by the plants for the purpose of describing the respiratory cycle.

The formation of water yields energy, which is used, with the aid of light energy, during the time when the store portion expands and the radiant portion contracts, to assist in the phosphorylation of adenosine diphosphate to the triphosphate compound.

When the store portion contracts the hydroxyl group may be accepted by the fat-soluble vitamins, especially the xanthophylls, as is evidenced by the changes these compounds undergo in the autumn and winter, the seasons in which the store portion is more contracted than expanded. In the autumn the xanthophylls increase and the chlorophylls diminish, to cause the leaves to change their green for a yellow colour. In the winter, when the store portion becomes more contracted than in the autumn, the xanthophylls become still more oxidised, causing the leaves to turn a brown colour and to fall to the ground.

## OTHER PLANT SYNTHESES

Although the assimilation of carbon dioxide, and its conversion into carbohydrates, is the most important of the syntheses of which the animate products are capable; nevertheless, there are others, only those of a vegetable nature are able to effect. The compounds which animals may require and only plants can synthesise, include the glycosides and fat-soluble vitamins in the store portion, and the water-soluble vitamins and the essential amino acids in the radiant portion. The synthesis of these compounds, although dependent on the description of the photosynthetic cycle, their synthesis does not form part of the description, hence the reason for the syntheses being considered under a separate heading.

Before considering the matter, the question arises. Why should plants, the more lowly evolved of the animate products, be given the tasks which might be expected to be more readily executed by the more highly evolved products, the animals? My answers are: first, that owing to the store portion of vegetable protein being relatively more contracted and the radiant portion relatively more expanded, and consequently more reactive, than those of animal protein, plants had a harder task to evolve from the beginning than ever became the lot of animals. And second, that having succeeded in executing the task, particularly in synthesising compounds animals find to be necessary for their evolution, better than might have been expected, the need never arose for animals to synthesise these compounds—their energy could be better expended in effecting a more efficient evolution.

All these compounds lend support to the first answer given. A property common to many of the compounds in the store portion and to the aromatic amino acids in the radiant portion; and, which is conspicuously absent in the store portion of animal protein, is the presence of a benzene ring. The compounds in the animal protein, which most consistently possess this ring, are the oestrogens; and these compounds are not wholly of store portion origin, having reached the portion by invasion from the radiant portion, through the attractive portion.

## THE GLYCOSIDES

The glycosides follow δ-carbon carbohydrates combining with an alcohol, and the compound formed undergoing dehydration. A relationship exists between the degree of contraction the store portion has been made to undergo, and the extent to which the glycoside is oxidised, dehydrated and is able to polymerise. Other compounds, whose ability to polymerise is influenced by the degree of contraction the portion to which they belong is made to undergo, are the nucleotides in the radiant portion. Attention is drawn to this point, as it may afford an explanation for the way in which viruses are formed when the radiant portion is made to over-contract, following the over-expansion "climate" had earlier made the store portion undergo. Should polymerisation be found to be the explanation, it would at the same time account for the unreactivity of viruses being due to the tying up of the catalysts in their formation; and, for the action displayed by vaccines, made from viruses, in preventing potential victims of the physico-chemical changes the protein is made to undergo in their formation, from becoming actual victims. What the vaccine does is to expand the over-contracted protein, the effect of which is to release the phosphate group from the nucleotides rather than the hydrogen from the water-soluble vitamins; and, permit it to travel to and fro between the store and radiant portions in the description of the pulsatory cycles.

## THE FAT-SOLUBLE VITAMINS

The fat-soluble vitamins include A, D, E and K. All being constituents of the store portion, they are connected with the main chemical change the portion undergoes; namely, oxidation; indeed, they are the chemical compounds which are most concerned with this change. As oxidation is a change which is associated closer with growth and the organs resulting from it than reduction, it becomes understandable how these vitamins have come to be connected with growth, and the tissues and the vascular and urogenital systems, which originated from the store portion.

Vitamin A is related to the carotenoids, but exactly how the compound came to be associated with the eyes, which originated from the attractive portion, remains a mystery. It is possible that when the precursor of the eyes, the pigmented eye-spot, made its first appearance in the *Algae*, it established a connexion with the store portion by means of a tail similar to the one the chloroplasts established with the portion. The pigmented eye-spot would appear to have originated from the attractive portion, in between the occasions when the nucleus and chloroplasts originated from the radiant portion; and, to have disappeared before the latter in the regression some of the *Algae* underwent. The disappearance of the pigmented eye-spot differs from that of the chloroplasts, in being retained in the gametes after having vanished in their parents. This difference may possibly be explained by the disappearance of the pigmented eye-spot having resulted from absorption, caused by the over-contraction the upper part of the attractive portion was made to undergo. Whilst the disappearance of the chloroplasts was more likely to have resulted from the over-expansion, the over-contraction of the store portion had caused the radiant portion to undergo. Much the same difficulty arises in the matter of the relationship vitamin D bears to the calcium metabolism, as arose in that vitamin A bears to the eyes, and as will be seen below in that vitamin E bears to the muscles and female genital system. Vitamin D is related to the steroid compounds, which are definitely concerned with the physical changes the store and radiant portions, particularly the junction between them, undergoes during the description of the pulsatory cycles. It is in this junction, as in the other two junctions, only to a lesser extent, that the calcium metabolism is most active. In this metabolism in the first junction, the calcium is reactive only when the store portion contracts and the radiant portion expands. Indeed, it is when the metal is released in abnormal quantities, which happens when the radiant portion is made to over-expand, that clotting is produced as a result. First, of the store portion having been made to over-expand; and second, of the radiant portion having been made to over-contract.

Vitamins E and K may be considered together, as both are closely related to the quinones, and to the chlorophylls' phytyl tail; and, the connexion they have with disease is in the store portion of the protein being over-contracted. In muscular dystrophy and sterility, the areas in the store portion from which the muscles and female genital system originated, are over-contracted. Haemorrhage is a result of the area in the store portion, from which the vascular system originated, becoming over-contracted, and any influence vitamin E may have in preventing its occurrence, is due to the action the preparation has in staving off the effect by expanding the area.

## THE WATER-SOLUBLE VITAMINS

The water-soluble vitamins include vitamin $B_1$, or thiamine; vitamin $B_2$ complex, which is composed of riboflavin, nicotinamide, folic acid and pantothenic acid; vitamin $B_6$, or pyridoxine; vitamin $B_{12}$, or cobalamine; vitamin C, or ascorbic acid; and vitamin H, or biotin.

Thiamine is composed of a pyrimidine and thiazol ring, which are connected at the so-called "positively charged" nitrogen atom of the latter. In its diphosphate form, this vitamin acts by breaking down the carboxyl groups in the description of the respiratory and citric acid cycles, into carbon dioxide and hydrogen, and becoming converted into active acetaldehyde in the process. The aldehyde group is formed at the carbon atom in the 2-position of the thiazol ring, from where it may be transferred to other compounds. The inability of thiamine to become connected in active acetylaldehyde may result in the development of polyneuritis in animals and *Beriberi* in man. The inability arises from the radiant portion having been made to expand and the manifestations from the attractive portion having been caused to undergo the opposite physico-chemical change.

Riboflavin and nicotinamide may be considered together, as the form in which they exist as catalysts is the same, and similarly the action the catalysts have and the manifestations which are produced when they are rendered inactive. Riboflavin is composed of a fused pteridine and benzene ring and of a ribityl side chain, whilst nicotinamide is a simple pyridine compound. Both exist as nucleotides when acting as catalysts; riboflavin in the form of flavin-adenine dinucleotide; and nicotinamide in the forms of nicotinamide-adenine dinucleotide and nicotinamide-adenine dinucleotide phosphate. All three nucleotides act by transferring hydrogen to the store portion, when the radiant portion expands. The transference is effected directly to the oxygen atom, to produce hydrogen peroxide with the assistance of flavin nucleotide.

In the case of the first nicotinamide nucleotide, the hydrogen is used for the production of water from oxygen in the description of the respiratory cycle; and in that of the second nucleotide for the synthesis of the hydrates in the description of the photosynthetic cycle. Hydrogen fails to be radiated to the store portion both when the radiant portion is over-contracted and over-expanded. It is only in the latter instance that manifestations are produced, and they are of a cutaneous, ophthalmic or nervous nature, depending on the area in the attractive portion which is made to over-contract. The radiant portion fails to radiate when it is over-contracted; owing to the hydrogen being rendered inactive, and when it is over-expanded the radiation is to the exterior rather than to the store portion. *Cheilosis* is the most common manifestation to be produced when the riboflavin nucleotide loses its catalytic action, and *Pellagra* when the pyridine nucleotides suffer the same fate.

Folic acid and cobalamin are alike in failing to act as catalysts, as a result of being unable to be absorbed; and, in their stead either a form of anaemia or thrombocytopenia is apt to be produced.

Folic acid is composed of pteroic acid and glutamic acid, and, in addition to the role its absence in man plays in the production of anaemia, the growth of micro-organisms appears to depend upon its presence; particularly upon the *p*-amino-benzoic which is also a constituent of the compound. Bacteria also depend for their growth upon the presence of cobalamine, whose ring differs from that of the porphyrins solely by failing a single methine group (=CH—). The role this cobalt compound plays in the haemin system has yet to be ascertained, but as similar compounds exist in which the cobalt kation is replaced by an anion, and all are found to be combined with nucleotides, it is obvious that at least one of the roles played by vitamin $B_{12}$ is in the maintenance of the acid-base, or reduction oxidation system.

Pantothenic acid, vitamin $B_6$, or pyridoxine, and biotin fall into a single group, as their catalytic action is more concerned with the metabolism of the amino acids, and their connexion with disease is exclusively with its cutaneous and not partly with its nervous and vascular manifestations, as is that of the other water-soluble vitamins. Pantothenic acid connects this group with the one just described, in being another water-soluble vitamin upon which micro-organisms depend for their growth. The compound is a dipeptide, consisting of dihydroxy-dimethylbutyric acid and alanine, and it combines with mercaptoethylamine to form pantetheine, which is the most important component of the equally important catalyst, active acetate. The importance of pantetheine, lies in its sulphydryl group (SH) being the active group of the catalyst, which, by activating acetates and higher fatty acids, enables carboxyl groups to be transferred. In the compound the acetyl residue ($CH_3CO$—) is bound to the free sulphydryl group, as may be seen in the action the catalyst has in converting choline into the acetyl compound.

Pyridoxine is a pyridine derivative which needs to be phosphorylated before being able to act as a catalyst, and whose action in this form is to transfer amino groups.

Biotin, sometimes known by the name of " vitamin H," is a cyclic derivative of urea in combination with a thiophane ring. In its catalytic form the compound, aided both by activated acetate and adenosine triphosphate, acts by transferring carboxyl groups.

Vitamin C, or ascorbic acid, is a carbohydrate which forms a constituent of the store portion, in the area where it is joined to the radiant portion, and where its action is influenced by whether the portion be expanded or contracted. When the portion expands the compound becomes active and undergoes dehydrogenation; whereas when it contracts the compound, having lost the hydrogen atoms of the two hydroxyl groups carried by the double bonded carbon atoms to the radiant portion, assumes its oxidised form. When the store portion is made to over-contract, which renders the vitamin inactive, scurvy is the manifestation of disease most likely to be produced.

## THE ESSENTIAL AMINO ACIDS

The essential amino acids include valine, leucine, isoleucine, lysine, methionine, threonine, phenylalanine, tyrosine and tryptophane.

Too little is known of the role valine, leucine, isoleucine and lysine play in the metabolism of the animate products, to warrant anything more being said about them. In the case of methionine the very reverse is the case, as not only is this amino acid one of the few to contain a sulphur atom in its composition but also it is the principal donor of the methyl group. The methyl group is attached to the sulphur atom and the transference of the group, which is mainly to an amino group, is effected by the compound combining with adenosine triphosphate.

Except for mentioning that threonine was the first of the group to be found essential, no more can be added to this amino acid than to the first four.

Phenylalanine, tyrosine and tryptophane may be considered together, as they are common in possessing an aromatic ring in their composition, which only the vegetable products are capable of synthesising. The first step in the synthesis is the formation of dehydroquinic acid, which results from the combination erythrose phosphate forms with pyruvic acid. Following the elimination of water, shikimic acid is formed, which proceeds to become phosphorylated in the 5-position. The compound then combines with phosphoenol pyruvic acid to form prephenic acid, which loses a phosphate group in the process. On losing further carbonic dioxide and water, prephenic acid becomes transformed to phenylpyruvic acid, which, upon transamination, becomes converted into phenylalanine. Tyrosine differs from phenylalanine in having an hydroxyl group in the *para*-position of the benzene ring. The oxygen in this group comes from the air, and the hydroxylation is effected by the catalytic action of the pyridine nucleotides. If through an inborn error of metabolism the hydroxylation fails to be effected, phenylketonuria may be produced, a condition which may be associated with feeble-mindedness and pigment deficiency. In addition to melanins, for which tyrosine is responsible, must be added thyroxine and adrenalin.

Tryptophane is formed from anthranilic acid, which supplies the aromatic ring, and the nitrogen for the indole ring. The importance of tryptophane lies in its being the amino acid responsible for the water-soluble vitamin, nicotinamide, which forms the basis of some of the most active of the catalysts, and for the compounds, serotonin and melatonin which form important constituents of the brain.

What is so remarkable about these essential compounds, to which must be added the highly unsaturated fatty acids to complete the story, which animals are incapable of synthesising, is that bacteria and fungi, in the order mentioned, are even better able to effect the synthesis than the more highly developed vegetable products. Mention is made of this point as, in my opinion, it lends support to the view, that the fungi are regressive forms of the *Algae* and the bacteria of the fungi, rather than it was from these forms that the more highly advanced plants evolved. And also that it affords an explanation for how symbiosis, saprophytism and parasitism became introduced; because despite the regression, a connexion between the various forms was never completely severed.

Two forms of regression occurred, the complete and the incomplete; and, the main interest about them, in the present context, is that the cycles which had the greater influence on the two forms were not the metabolic but rather the pulsatory—showing that in spite of the former playing the greater role in the evolution of the products, the role could never have been played had it not been for the latter. The complete form of regression occurred when the store portion of the vegetable products became so over-contracted as to cause the radiant portion to over-expand, to the point where the chloroplasts, which were the second organs to originate from the portion (the nucleus having been the first), were caused to vanish. And consequently the attractive portion to over-contract, in the process of which the pigmented eye-spot became obliterated through absorption. The result was the regression of the *Algae* into fungi and the latter into bacilli, sub-bacilli and cocci.

In the incomplete form of regression the store portion was less contracted, and, the over-expansion, the radiant portion was made to undergo was reflected onto the male sex, which was the third organ to originate from the portion, to cause it to regress rather than to vanish. The result was the alternation of generations, a form of reproduction. Once these forms of regression were set in motion, the movement appears never to have come to a halt. This continuity is shown, in the complete form, by the more pathogenic a bacillus becomes the further it regresses, and, in the incomplete form, by the male and female gametes in the fucoids having become replaced by the pollen-grains and the embryo-sac in the seed plants. In the incomplete form the female sex never suffered the regression of the male sex, a difference it owes to having been more successful in making the store portion its home than its partner.

These two forms of regression can be traced to plants having been forced to take up a static position, and one which made life still more difficult for them once they left the water to come to live on land, where drier conditions prevail.

Animals were saved from suffering the same fate, first, through being able to move about in search of their food, and second, in finding it in the plants in which the metabolic cycles had been adequately described. Not only did the need arise for the store portion to over-contract and the radiant portion to over-expand, but also the attractive portion was kept intact, an event which enabled the animal products to replace the radial symmetry of the vegetable products by the bilateral symmetry. In other words, instead of the radiant being practically the only portion with which the store portion could establish a connexion in the vegetable protein; in the animal protein, not only was the store portion made to unite with the attractive portion in addition, but also something of the nature of a competition arose between the two reactive portions with which the unreactive portion was to become connected.

The radiant being practically the sole reactive portion with which the unreactive portion is able to be connected, may provide the explanation for the way in which leaves are found to be spirally set around the stem in plants. The spiral form is that assumed by the protein in the junction between the store and radiant portions and the organs which originated from it, as is evidenced in fibrin, *Spirilla, Spirochaetae,* etc. And it is possible that this form arose from this junction being the one in which the physical changes of expansion and contraction became reversed in the pulsatory cycles " activity " describes in the product.

## FERMENTATION CYCLE

*(Anaerobic glycosis)*

This cycle is the one in which one of the more regressive of the vegetable products, a fungus, or a bacillus, is employed to break down the carbohydrates in the store portion of one of the seed plants into alcohol. The complex in the radiant portion of the seed plant held to be responsible for the degradation of the starch, which is the material formed by the sugars on reaching their apex of contraction, is called " β-amylase." Apparently, the order in which the amino acids are arranged in this complex has yet to be ascertained. The first step to be taken in the description of the cycle, is surmised to be a union between the starch and the β-amylase. In the event of a union being established the question arises. What prompted it? My answer is the expansion the store portion, in which the substrate is situated, and the contraction the radiant portion, in which the complex is situated, undergo in the description of the pulsatory cycles. I picture the union as resulting from the relationship that became established between the formation of the starch, and the sequence into which the amino acids in the complex became arranged, at the time, in the evolution of the products. And from the urge the difference between the interacting products gave them to reach the state of stable equilibrium. The relationship and difference is not only those that exist between expansion and contraction, but also, in the present case, those that exist between benefactor and attacker. A fungus, or bacillus, may either supply a seed plant with food, or become parasitic upon it. The result of the union is the splitting off of maltose from the ends of the starch. The splitting occurs at the site of the oxygen atom, which binds the carbon atoms of the maltose compounds in the 2- and 5-positions. Following the splitting, a phosphate group becomes attached to this oxygen; and, there would appear to be three sources from which this phosphate group may be derived. It may be stored in the starch, found free in the store portion, or donated by the nucleotide, adenosine triphosphate, from the radiant portion, at the time the portion expands.

In the second step of the cycle the maltose is split by another complex, with the addition of adenosine triphosphate, into two compounds of hexose-monophosphate, each of which, with the further aid of a complex and adenosine triphosphate, becomes converted into hexose-diphosphate. In this diphosphate compound, the second phosphate group replaces the hydrogen in the hydroxyl part of the hydroxy-methyl group in the 1-position. The replaced hydrogen is accepted by the catalyst nicotinamide-adenine nucleotide in the radiant portion, at the time the portion contracts.

With the assistance of another complex and a further supply of adenosine triphosphate, the splitting is carried a stage further; each hexose-diphosphate compound is split into one compound each of phosphoglyceraldehyde and phospho-dihydroxyacetone.

Both of these compounds become converted into phosphoglyceric acid, the former by means of dehydrogenation and the latter of phosphorylation. In the former the complex responsible for the reaction is one containing an active sulphydryl group (—SH); whilst in the latter the sulphydryl group is replaced by a disulphide group. In the case of phosphoglyceraldehyde, the complex unites with the carbon atom of the aldehyde group to result in the freeing of hydrogen, which is accepted by the nicotinamide-adenine nucleotide in the radiant portion. In the case of phospho-dihydroxyacetone, the carbonyl group is oxidised by the complex and phosphorylated by adenosine triphosphate, when the radiant portion expands. When the phosphate group has done its work in carboxylating the carbonyl group, the phosphate group is accepted by adenosine diphosphate, which becomes transformed into the triphosphate compound on the radiant portion contracting.

Phospho-glyceric acid is degraded into phospho-pyruvic acid, and the latter into pyruvic acid. These two forms of degradation depend upon the action of complexes, and in the processes the phosphate group is released and replaced by hydrogen. The released phosphate group is accepted by adenosine diphosphate, which becomes transformed into the triphosphate compound as the radiant portion contracts; whilst the hydrogen is donated by the reduced nicotinamide-adenine dinucleotide as the radiant portion expands.

In the breakdown of pyruvic acid into acetaldehyde, the compound undergoes decarboxylation, in the process of which both magnesium and thiamine pyrophosphate act as catalysts. The final change into alcohol is reached by nicotinamide-adenine dinucleotide donating hydrogen to the acetaldehyde.

Although the energy produced in the description of the cycle suffices for the changes the compounds involved are made to undergo, insufficient is stored in a chemical form to enable the changes to be repeated in the reverse order, as is rendered possible by the presence of oxygen in the description of the other secondary cycles. Thus explaining why alcohol, when once formed, remains unchanged for a more or less indefinite period.

## AEROBIC GLYCOLYSIS

In this form of carbohydrate degradation exactly the same steps are taken as in the anaerobic form, right up to the point where pyruvic acid is formed. The difference which takes place at this point, is that pyruvic acid is broken down into carbon dioxide and water, instead of through acetaldehyde into alcohol. In the aerobic breakdown three courses are open to be taken; oxidative carboxylation, reductive carboxylation and transamination. In the first course, in which the catalyst thiamine pyrophosphate interacts with pyruvic acid, carbon dioxide is released forthwith; and, what remains of the acid combines with the catalyst to form what is known as " active aldehyde ". This name is derived from the role the combination plays in the formation of the catalyst, which is responsible for the transference of the carboxyl groups in the amino acid metabolism. The catalyst in question is composed of pantoic acid, β-alanine and mercaptoethylamine. In the second course, pyruvic acid, in company with carbon dioxide and catalysed by nicotinamide-adenine dinucleotide phosphate, is made to enter the citric acid cycle, where it becomes transformed into malic acid. This malic acid, following its immediate dehydrogenation by the catalyst nicotinamide-adenine dinucleotide, becomes changed into oxaloacetic acid, the compound which sparks off, with the aid of the activated acetate catalyst, the description of the citric acid cycle. In the third course, pyruvic acid is transformed into the amino acid alanine, which becomes, as mentioned above, a constituent of the catalyst, active aldehyde.

The carbohydrates are not the only chemical components of protein whose final degradation is the production of carbon dioxide and water, as these products are also those into which the fats and amino acids are finally broken down.

The main source of the carbon dioxide stems from the decarboxylation which goes on in the description of the citric acid cycle; and, of the water from the oxidation which occurs in that of the respiratory cycle.

" Respiratory " is a name that is given to the cycles in which the carbohydrates are the chemical compounds to be broken down. The similarity between the citric acid and respiratory cycles lies in their being described in the store portion of the protein, the chemical compounds most responsible for their description being those which have their home in the radiant portion. Not only does this relationship, that has been shown to exist between the two portions, confirm the store portion to be the primary portion but, in addition, that the manner in which the pulsatory cycles are described in protein must have an influence upon the manner in which the metabolic cycles are described in the product. Catabolism is not the only change the chemical compounds undergo in the description of the citric acid cycle, anabolism is another. Indeed, in the anabolic half, compounds are formed which initiate the synthesis, not only of compounds which become components of the store portion but also of those which become components of the radiant portion. This point confirms still further the importance of the store portion, and shows the cycle to be the most important of the metabolic cycles. In actual fact the citric acid cycle is the one on which the metabolic cycles revolve. Finally, the catalyst which plays the greatest role in the syntheses, for which the description of the citric acid cycle is responsible, is active acetate.

## THE CITRIC ACID CYCLE

The first step to be taken in the description of this cycle is by oxaloacetic acid, which can arise either from aspartic acid or pyruvic acid. With the aid of activated acetate, oxaloacetic acid is transformed into citric acid. The citric acid in two further steps becomes changed into the iso-compound, which on being catalysed by nicotinamide-adenine dinucleotide phosphate, undergoes dehydrogenation to become converted into oxalosuccinic acid; and later decarboxylation to result in the production of the important compound α-ketoglutaric acid. The importance of this acid, in the present context, lies in its formation and transformation into succinic acid being the means by which carbon dioxide is produced and released from the cycle. In the formation the process is reversible, but not so in the transformation, in which, in addition to the decarboxylation, dehydrogenation occurs along with the loss of a carbon atom. Consequently, in the transformation of α-ketoglutaric acid into a succinic acid, a complicated process is involved. In the decarboxylation alone the catalysts thiamine pyrophosphate, lipoic acid and the combined adenosine diphosphate and pantetheine phosphate play their respective parts. In the dehydrogenation the catalyst nicotinamide-adenine diphosphate becomes changed into its reduced form, and until succinic acid is actually formed, the combined catalyst continues to remain active. With the aid of the catalyst flavin-adenine dinucleotide succinic acid is dehydrogenated into fumaric acid, the catalyst being reduced in the process.

In the final two steps fumaric acid, with the addition of water and catalysed by a complex of unknown nature, is transformed into malic acid. Whilst in the case of the latter acid its hydroxyl group is dehydrogenated; the hydrogen atoms being accepted by the catalyst nicotinamide-adenine dinucleotide.

In the citric acid cycle, which is described in conjunction with the respiratory group of cycles, water is taken up, and activated acetate is broken down into carbon dioxide and hydrogen, the latter of which is accepted by the pyridine nucleotides in the radiant portion, on the occasions the portion contracts, and in the process of which the catalysts become reduced. Although degradation is the major feature of the description of the cycle, there exists two minor ones to be considered. The first, is the cycle's acceptance of the C2 compounds, which are released in the break down of the carbohydrates, fats and amino acids; and the second, is the synthesis of glucose, amino acids, particularly aspartic and glutanic, and porphyrin.

## THE RESYNTHESIS OF GLUCOSE AND OTHER SYNTHESES

The basic chemical compound from which glucose is resynthesised is phosphoenol pyruvic acid. This compound is formed directly from oxaloacetic acid by the latter being decarboxylated and phosphorylated by inosine triphosphate; and indirectly from malic, lactic and the amino acids. The indirect formation from malic acid, is effected by the compound undergoing dehydrogenation when catalysed by nicotinamide-adenine dinucleotide, in the course of the store portion expanding. The hydrogen atoms are accepted by the catalyst, which undergoes reduction as the radiant portion contracts. In the indirect formation from lactic acid, which is formed when glycolysis is made to take place anaerobically. The compound is transformed into pyruvic acid in the same way as is malic acid into oxaloacetic acid; and, the pyruvic acid into malic acid by the process of reductive carboxylation. The catalyst in this process is the reduced form of nicotinamide-adenine dinucleotide phosphate, which loses its hydrogen atoms as the store portion contracts and the radiant portion expands. Once phosphoenol pyruvic acid is formed, the compound, with the aid of adenosine triphosphate, undergoes the reverse changes of those which take place in glycolysis.

Transamination is the process by means of which the organic acids are most frequently produced from the amino acids; and, as might be expected, the one produced is determined by the amino acid from which it has been obtained. Only those amino acids, which can be transaminated, either to pyruvic acid or one of the two dicarboxylic acids are available for glucogenesis. In this process of transamination, following the union between the amino acid and pyridoxal phosphate, the hydrogen attached to the carbon atom becomes double bonded, and the combination is split into an organic acid and pyridoxamine phosphate. Oxidative deamination is another process by means of which organic acids may be produced from amino acids. In this process two hydrogen atoms are eliminated and accepted by the catalyst, flavin-adenine dinucleotide, and, the amino acid becomes converted into an imino acid, which is finally hydrolysed to form ammonia and an organic acid. The fate of the ammonia is to be consumed in the formation of urea.

## THE FATS

Fats are combinations of fatty acids and glycerol. The fatty acids are built up most readily from the 2-carbon compounds, the simplest one of which is acetic acid. In the first step of the building up of these carbon compounds, acetic acid combines with adenosine diphosphate and pantetheine phosphate to form the all important catalyst, activated acetate. In the second step this activated acetate, with the addition of a carboxyl group, obtained from the water-soluble vitamin, biotin, and catalysed by adenosine triphosphate, becomes transformed into malonyl active acetate. In the third step this malonyl active acetate combines with an active acetate compound to form a β-keto acid, in the process of which decarboxylation occurs. The carboxyl group removed is the one that is employed in the formation of malonyl active acetate.

Following hydrogenation and phosphorylation by the reduced form of nicotinamide-adenine dinucleotide phosphate, aided by a flavin catalyst, the β-keto acid becomes changed into butyryl active acetate. This compound on combining with malonyl active acetate, which undergoes decarboxylation in the process, results in the formation of the next fatty acid in the scale and so on.

The glycerol is a reduction product of the dihydroxyacetone phosphate, which is formed in the degradation of the carbohydrates, the catalyst being the reduced form of nicotinamide-adenine dinucleotide. In addition to being a component of the fats, glycerol phosphate is the basic constituent of the phosphatides, which, in combination with calcium, have their homes in the junctions between the three portions of the protein, rather than in the portions themselves. This difference in the position occupied by these compounds, accounts for their presence in the nucleus, which originated from the first junction, and its counterpart, the sense-organ mechanism, which originated from the third junction and in the ductless glands, the evolutionary products of the specialised sense-organs which originated from all three junctions. And for their absence in the ductless glands, the evolutionary products of the generalised sense-organs, which originated from the portions.

The phosphatide is synthesised from glycerol and phosphorylated by the compound combining with one active acetate.

The radical of the latter becomes part of the phosphatidic acid, which is formed with the aid of the catalyst cytidine diphosphate, leaving behind in the process active acetate in its inactive form and a phosphate group.

## THE STEROIDS

In the synthesis of the steroids, in which, for the third time, active acetate is found to be the principal catalyst, the basic chemical compound is the 5-carbon compound, isoprene. The first step in the synthesis is taken by the active acetate combining with the aceto-acetyl compound, which is formed by two compounds of the catalyst uniting to form β-hydroxy-β-methylglutaric acid. Which on being hydrogenated by the reduced form of nicotinamide-adenine dinucleotide phosphate, results in the formation of mevalonic acid. In the second step mevalonic acid, following its phosphorylation with adenosine triphosphate and its dehydroxylation and decarboxylation, with a further supply of the same catalyst, becomes converted into isopentenyl pyrophosphate. In the third step this active isoprene compound undergoes a series of changes, amongst which polymerisation and condensation occur, to result in the production of a compound containing thirty carbon and fifty hydrogen atoms, and which is called squalene. This compound is composed of two symmetrical halves both of which, following hydroxylation, cyclisise and undergo further changes to result in the production of the sterols. Forming as it does the starting material for the synthesis of the steroids, cholesterol must be considered as being the principal sterol to have been produced.

## THE NITROGEN COMPOUNDS

Leaving the synthesis of the carbohydrates, fats and steroids, the compounds which constitute the basis of the store portion, to pass to that of those constituting the basis of the radiant and attractive portions, attention needs to be drawn to an all-important difference between the two. The difference is the presence of nitrogen in every organic component of the two reactive portions, compared with its absence in every component of the unreactive portion.

Although the nitrogen compounds are those which are most responsible for keeping the radiant and attractive portions reactive, the reactivity cannot be ascribed to them, as the portions were reactive before nitrogen appeared. The reactivity arose from the ability of the radiant portion to accept the " activity," which is liberated from the store portion, during expansion, when the portion contracts, in the pulsatory cycles which " activity " describes in the products of its condensation.

All three portions are reactive when they expand. The store portion is rendered unreactive on contracting, and prior to being rendered reactive on expanding, the portion needs to have " activity " subtracted from it. The acceptance of this activity by the radiant portion, prevented the " activity " liberated from the store portion from becoming dissipated, without affecting " activity " being radiated in the portion expanding.

In order to prevent the loss of " activity " from radiation, need arose for " activity " being attracted, which is the function exhibited by the attractive portion on expanding. The function of attraction being the opposite to the one of radiation, gives its exhibition an independence that of radiation does not possess. This difference accounts for the contraction the attractive portion undergoes being indivisible from that the radiant portion undergoes. The nitrogenous compounds responsible for all these physico-chemical changes, which the reactive portions of protein undergo in the description of the pulsatory cycles in the product, are the various forms capable of being adopted by ammonia. From this the conclusion may be drawn that ammonia is the form in which nitrogen entered the compound, and into which any other form on entering needs to be reduced prior to becoming operative. When the store portion of protein expands, in addition to " activity," hydrogen is liberated from the compounds constituting its basis, and first from the carbohydrates. In the contraction the radiant portion undergoes in response to this expansion, the liberated hydrogen is accepted by the amino group, in the water-soluble component of the vitamin catalyst. As the contraction extends in the radiant portion, dehydroxylation of the carboxyl group in the amino acids occurs, to result in the amino group uniting with the carbonyl group to form the peptide link. In the expansion the attractive portion undergoes, following this contraction, the amino group suffers dehydrogenation, a change which enables it to unite with a methyl group, which has been donated by the amino acid, methionine. As the expansion of the attractive portion extends, the methylation of the amino group increases, to result in the independence enjoyed by the attractive portion when it expands, being replaced by interdependence; and, the attraction of " activity " by expansion of the store portion. The first of these results follows the nitrogen becoming pentavalent, a change which causes it to exhibit predominantly the function of storing; and the second, by making the methyl groups independent of the amino group. So long as the methyl group remains part of the amino group, the amino-methyl group is the group that is independent; and, in the state that enables it to exhibit the function of attraction.

The exhibition of the function of storing, which is initiated by the amino group becoming pentavalent, in the junction between the attractive and store portions, extends to involve the store portion, which then proceeds to contract. The contraction is unreactive due; in part to the compounds most affected being the steroids, which acquired the property of the amino group when it became pentavalent; and in part to the increasing oxidation the compounds undergo following the liberation of the hydrogen, which occurs when the store portion expands. This increasing oxidation receives its first check in the junction between the store and radiant portions, where the carbonyl group of the steroids becomes replaced by an aldehyde group, the hydrogen of which begins to render the contractile action of the compound reactive.

In the expansion the radiant portion undergoes in the second half of the description of the pulsatory cycles, the hydrogen accepted by the amino group in the first half is radiated to the store portion. This radiation has a two-fold effect; the first of which is to stop the expansion; and the second to reactivate the unreactive contraction. This double action is repeated, in a modified way, by the metallic atoms, which are attached to the amino group in the porphyrins; and, the hydrogen atom in the carboxyl group of the amino acids. The modified way in which the metallic atoms act is to increase the reactivity; first, of the radiant portion; and second, of the store portion when they contract. Whilst that in which the hydrogen atom of the carboxyl group acts, is to increase still further the reactivity of the radiant portion when it contracts. Any hydrogen not utilised by the portion is radiated from it to the exterior. On account of the independence the attractive portion gains when it expands, little can be said about the portion when it contracts, as this part of the portion is indivisible from the corresponding part of the radiant portion. What appears to be the main characteristic of the attractive portion when it contracts, is the use that is made of the peptide group in its reactive contractile form. A good example of the use of this group is in the compound, melatonin, which is an active principle of the pineal gland, and one of those to which the gland owes its contractile action.

In these other syntheses the first compounds to be considered are the nucleic acids. But there is something more to be said about these compounds and their relatives, before attention is paid to the way in the components of the complexes, for which they are responsible, are built up and those containing nitrogen are synthesised. Stress is placed upon the phrase, components of the complexes, as those of the nucleic acids may, one day, be discovered as having been the first to have come to be composed of so many different kinds of chemical compounds. The nucleic acids are composed of calcium, phosphoric acid, ribose, the pyrimidines, cytosine, uracil and thymine and the purines, adenine, hypoxanthine and guanine.

The complexes exist in the two forms of the desoxyribose nucleic acids and the ribose nucleic acids. In these two forms the components of the complexes are differently arranged; the pyrimidine, uracil, being found almost exclusively in the first form, and thymine, in the second form. Other characteristics of the complexes are their ability to polymerise in their original form; to combine with the water-soluble vitamins to form more reactive complexes, or catalysts, and to become the foundation; first, of the nucleus; and second, of the sexes. The divisibility of the sexes derives from that of the two forms of the nucleic acids.

The prophyrins resemble the nucleic acids in being composed of metallic atoms and nitrogenous chemical compounds; in being able to polymerise; and in being responsible for the genesis of such chemical compounds as the cytochromes, and organs as the chloroplasts and the red blood corpuscles. The amino acids are similarly able to polymerise, and to be responsible for the development of organs, such as the leucocytes, pulmonary and portal systems, the parathyroid, thymus, islets of Langerhans and thyroid glands. But, in the first place, the complexes are composed of different amino acids; in the second place, these amino acids are arranged in different sequences; and in the third place, the organs are singular rather than multiple as are the nucleus, the sexes and the chloroplasts. The nucleic acids owe their uniqueness to having originated from what came to be the junction between the store and radiant portions of the protein; an event which made the first to be the most important of the three junctions.

I venture to conclude, it was due to the indivisibility of the two portions at the time of the formation of the first junction, that the nucleic acids and sexes owe their divisions into the reduced and oxidised and male and female forms. Both nucleic acid and sex, in their pristine forms, originated from the radiant portion, in what was to become desoxyribose nucleic acid and the male sex; and contraction of the portion having been more responsible for the origin than the expansion. Owing to the genesis of these forms being repetitive, which allowed some to enter the store portion in which expansion is more reactive than contraction, the desoxyribose nucleic became transformed into ribose nucleic acid and the male into the female sex. This invasion of the store portion by nucleic acid and sex wrought other than chemical changes on the invaders. In the first place, it reversed the repetitiveness of the products; and in the second place, it caused the invaders to become controllers of the results arising from the leadership of the original products—the physico-chemical relationship established between the two pairs having been the factor, which necessitated their acting oppositely, when uniting, from the very beginning. The first is best exemplified by the singularity of the female gamete when compared with the multiplicity of the male gamete, and the second by the female being the only sex capable of producing another sex, which must invariably be of its own kind, hence the name " parthenogenesis " by which the process is known. The lead given to the store and the attractive portions by the radiant portion, in the description of the secondary cycles explains, not only the role the polypeptides play in what have come to be called " ferments," or " enzymes." But also that, particularly the nucleotides and the water-soluble vitamins, play in what have come to be called " co-ferments " or " co-enzymes," for want of the knowledge of how these chemical products came to take the lead.

The influence the manner in which the store portion behaves, i.e. the extent to which it expands and contracts in the pulsatory cycles, has on the results of the leadership, is in no way better shown than in that it exerts on the polymerisation the nucleotides may be made to undergo to produce what have come to be called "viruses." A direct relationship exists between the principal abnormal physico-chemical change the store portion is made to undergo, and the kind of virus which is formed, in the acute stage of disease. Similarly the cytoplasm of the epithelial cells; and this along with the nucleus of the nerve cells and the kind of inclusion body which is formed, in the acute stage. And the changes the nucleolus in the nucleus of the cells of the affected organ may be made to undergo in the morbid condition of cancer. In all these instances the polymerisation would appear to be affected by an hydroxy-methyl group, which is attached to the carbon atom in the 5-position of one or more of the pyrimidine and purine compounds in the nucleic acids.

## THE NUCLEOTIDES

The components of the nucleotides whose synthesis comes under consideration are the pyrimidines, the purines and the water-soluble vitamins.

The compound needing to be formed, before combining with ribose, to produce the pyrimidine nucleotides is orotic acid. This acid is the dehydrogenated product of an open ring compound, which has resulted from the combination of the dicarboxylic amino acid, aspartic acid, and carbonyl phosphate. Following the decarboxylation of the nucleotide produced by the orotic acid combining with a phosphoribosyl pyrophosphate, uridine triphosphate is formed, the compound from which the other pyrimidines are derived. In the case of the purines, the ribose phosphate is prepared first and in the form of phosphoribosyl pyrophosphate. After obtaining an amino group from glutaminic acid, the compound combines with glycine to form the amide which constitutes the central piece of the purine ring. Following the addition of an aldehyde and an amino group the imidazole ring becomes closed. With the entry into the compound of carbonic acid, an amino group obtained from aspartic acid and an aldehyde group, inosinic acid is formed, from which the two purine nucleotides, adenylic acid and guanylic acid are derived.

Of all the nucleotides, which are synthesised in protein, none is more important than adenylic acid, owing to its being, in its triphosphate form, the first of its kind to act as a catalyst. The catalytic action of the compound is exemplified in its ability to transfer one or two phosphate groups from the radiant to the store portion, for the dual purpose of placing a check on the expansion the store portion undergoes, and of reactivating the portion after contraction has rendered it unreactive. Although other nucleotides are able to transfer other groups from the radiant to the store portion, it is not until the compounds combined with the water-soluble vitamins, which are derived from the amino acids, that catalysis is seen to be the important function of the radiant portion it is. And, most particularly, in the transference of the simplest of all the atoms, hydrogen. Here again, amongst some of the first of such complexes to be formed are those following the combination adenine dinucleotide and adenine dinucleotide phosphate form with nicotinamide, which is derived from the amino acid, tryptophane. In these compounds the hydrogen, which is liberated from the store portion as it expands, is accepted by the nicotinamide as the radiant portion contracts. In the process the pyridine ring undergoes reduction, loses one of its double bonds and the nitrogen is deprived of its so-called " positive charge." In this state the nicotinamide remains until the radiant portion expands, when the hydrogen is radiated to the contracted store portion for the purpose of rendering the portion reactive. To be more specific, the hydrogen, liberated from the reduced form of nicotinamide adenine dinucleotide, is employed for the purpose of synthesising adenosine triphosphate; whilst that liberated from nicotinamide adenine dinucleotide phosphate, is used for the synthesis of the carbohydrates from which the hydrogen was originally obtained. Attention is drawn to the different actions of these two catalysts; first, to emphasise the priority the phosphate group has over the hydrogen atom; and second, to suggest that the inclusion of this group in the so-called " antibiotics " might enhance their therapeutic action, which is none other than to normalise the abnormal physico-chemical changes the store portion is made to undergo in the acute stage of disease, totally irrespective of their cause. Another water-soluble vitamin to combine with adenine dinucleotide, to form an important catalyst, is riboflavin. This vitamin is composed of a combined pteridine and benzene ring, to which is attached as a side chain, a derivative of ribitol. The action of this catalyst is to assist the transport of the iron in its reduced form from haemin in the radiant portion, through the cytochromes, to become oxidised in the store portion.

Other water-soluble vitamins and amino acids themselves combine with nucleotides, for the purpose of transferring groups from the radiant portion both to the store and attractive portions. Two of the best known are the roles pantetheine phosphate and adenosine diphosphate play in the transference of carboxylic acid groups; and adenosylmethionine in the transference of methyl groups. This same complex of adenine, ribose and methionine, with the addition of guanidinoacetic acid, is responsible for the production of creatine.

In all these and other unmentioned reactions the nucleotides, headed by adenine triphosphate, act as catalysts; showing, that in addition to facilitating the leadership the radiant portion has over the other two portions, catalysts assist at the same time to energise the metabolism of the portion of which they have become the most essential components.

## THE PORPHYRINS

These compounds consist of four pyrrole rings, which are connected by methine groups (=CH−). Each pyrrole ring is composed of a combination of two compounds of amino-laevulinic acid, which has its origin in the union succinic acid establishes with glycine. The catalyst that came to hasten these reactions was active acetate, and its action resulted in more than one type of porphyrin being formed. The type forming the basis of chlorophyll, haemoglobin and the cytochrome is characterised by having a propionic acid side chain attached to the carbon atom in the 7-position, and an acetic acid side chain to the carbon atom in the 8-position of the fourth ring. The differences between these three types of porphyrins, lie in the different changes these and other side chains in the completed products undergo; and in the magnesium in chlorophyll being replaced by iron in haemoglobin and the cytochromes. The iron, similar to the magnesium is to be found in its divalent form, and in this form it remains until the metal has been transported, by the cytochromes and other catalysts, to the store portion, when it becomes trivalent. The only occasion on which the iron in the haemoglobin becomes trivalent is when it is converted in methaemoglobin, a change which occurs when the over-contracted radiant portion causes the blood corpuscles to over-expand.

The cytochrome about which most is known is the "c" variety. An interesting feature concerning this variety, is not only that the polypeptide (enzyme) is partially known from which the cytochrome (catalyst) is derived, but also how the two are combined to form a single product. In this product the combination is effected by the third group in the two compounds of cysteine in the polypeptide, forming an ether bond with what was originally the propionic group attached to the carbon atom in the first and fourth ring of the porphyrin compound.

In view of the lead the radiant portion of protein has gained over all three portions in the metabolism the product undergoes. And the specificity that has consequently arisen between the combinations which have been formed in the portion, between the polypeptides and the compounds which have originated from them, occurring coincidently with the combinations which have been formed in the other two portions, particularly the store portion, it is obvious that once the relationship between these combinations is established, the present trial and error method of treating disease will be able to be replaced by a truly scientific one. Stress is placed upon a knowledge of the relationships existing between the combinations which are formed in the store and radiant portions, in this order. For the reason that the store is the first portion to be attacked in disease; and that the physico-chemical changes it is made to undergo influence the lead which is given by the radiant portion.

The points raised above are well illustrated in the similarity which exists between the primary abnormal physico-chemical change the store portion of the protein is made to undergo to produce the acute, or seasonal, manifestations of disease and leucaemia; and, between the chemotherapeutic preparations which are employed to combat them. The primary abnormal physico-chemical change is over-expansion; and, the therapeutic preparations are either of a nucleotide or water-soluble vitamin nature. These compounds owe their therapeutic action to supplying the over-expanded store portion with the hydrogen needed to reactivate it, and of which the portion is being deprived through the over-contraction, the over-expansion of the store portion has caused the radiant portion to undergo. The only difference between the acute, or seasonal, manifestations and leucaemia, are that the over-expansion it is made to undergo, occurs in the acute stage of disease in the one and in the chronic in the other. And that the effect of this difference is to cause the radiant portion of the protein to over-contract in the first; and the nucleus, especially the nucleolus, of the mononuclear leucocytes in the second. Finally, these differences can be traced to the abnormal physico-chemical changes, which the store portion has been made to undergo in the interval between the acute and chronic stages.

In spite of the beneficent action these substances may have in the treatment of the above mentioned manifestations, plenty of room exists for an amplification of their action, particularly in the treatment of the manifestations that are caused to arise in the chronic stage. The factor occasioning this need to become apparent, and more so in the chronic than the acute stage of disease, is that the modern chemotherapeutic preparations act as aids to the leadership possessed by the radiant portion, rather than as the leadership itself, the latter of which becomes more urgently needed in the chronic than in the acute stage. This difference arises from the protein being the product in which the over-contraction occurs in the acute stage; and the first organ (the nucleus) which originated from it (the junction between the store and radiant portions) in the chronic stage. The leadership resides in the polypeptides, from which all the aids may be said to have had their origin. And further, the leadership lies not in the different amino acids of which the polypeptides are composed, but in the sequence in which these compounds are arranged. Indeed, the sequence has reached the level of importance which makes it both specific and genetic. Another reason for the therapeutic action of the catalysts being more efficient in the acute stage, and the polypeptide compounds responsible for the leadership in the chronic, lies in the main action of the former being to halt the over-expansion of the store portion, and of the latter to expand the over-contracted radiant portion. Opposites are needed to produce the best results in the first of these physico-chemical changes; whilst in the second their places are better taken by similars. This difference is well exemplified in the inability of the antibiotics to expand the viruses; and in the cortisone preparations having proved themselves to be the compounds best able to expand the over-contracted store portion. The cortisone preparations are of a steroid nature; and, they are the compounds to become most involved in the over-contraction the store portion is made to undergo in the subacute stage of disease.

## THE AMINO ACIDS

My reasons for interrupting the discussion on syntheses to enter into that on therapeutics are the following: The amino acids were the next and last compounds to be considered. About as much is known of these compounds individually as is required to be known, of those which are and are not essential; of many of the di- and tripeptide compounds they form; and, even of a few of those of a polypeptide nature which have only recently been unravelled. But it is not until the sequence in which the amino acids are arranged in the polypeptides, constituting the basis of the radiant and attractive portions, and of the organs which have originated from them, have been discovered, that any further fundamental progress in this and the therapeutic fields will be able to be registered. To round off the discussion, it is my present suspicion, that if ever a chemotherapeutic preparation be found to be of value in cancer, it will need to be similar to one of the polypeptides constituting the basis of the over-contracted area in the radiant portion from which the nucleus, especially its nucleolus, originated.

## THE REPRODUCTIVE CYCLE

THIS cycle resembles the other secondary cycles in the lead the radiant has over the attractive and store portions, and in which the results of the lead are influenced by the physico-chemical changes the store portion is made to undergo in the description of the pulsatory cycles. But it differs in two important respects; the first being, that the physico-chemical changes the store portion is made to undergo in the description of the pulsatory cycles have a greater influence on the lead given by the radiant portion, than is the case in the description of the other secondary cycles. The second being, that organs, in addition to atoms and chemical compounds, play a role in the description of the reproductive cycle which is not the case in that of the other secondary cycles. Indeed, these differences arise from the appearance of organs in the description of the reproductive cycle, and from first one and second the other of the two pairs of organs, after originating from the radiant portion, having come to make their home in the store portion.

The first pair of organs to appear were the two nuclei. These two became paired, as a result of the first one to originate from the radiant portion having migrated directly into the store portion in which it made its permanent home and acquired the properties of the portion. These properties include contraction, oxidation and unreactivity. The migration caused a second nucleus to originate from the radiant portion and one which was capable of reproducing itself, which its forerunner was not. Whether the migrating nucleus was formed by a process of replication or duplication is unknown, owing to the way in which the remaining nucleus reproduces itself, has yet to be determined. The nucleus, or nuclei, remaining in the radiant portion naturally acquired the properties of the portion, which include expansion, reduction and reactivity.

The second pair of organs to appear were the female and the male gametes. The female gamete followed the path of the first nucleus into the store portion, but indirectly, through the attractive portion, instead of directly from the radiant portion.

Although the male gamete continued to originate from the radiant portion and to retain the properties of the portion, the store was the portion in which the gamete ultimately made its home. Long before the pairs of nuclei and gametes had reached the above-mentioned level of evolution, the nucleus had become not only a constituent of every cell, but also the main one. The nucleus became the main constituent through having been able to retain the properties of the radiant portion, in addition to acquiring those of the store portion.

As a result of the many cells of which the animate products became composed, these dual properties came to be acquired by the products themselves. And first by the unicellular flagellated vegetable and animal products, which, on account of the properties being opposite, were enabled to reproduce, or propagate. The flagellates are unicellular cells composed of a nucleus, chloroplasts, or chromoplasts, a pigmented eye-spot and one or more flagella. The chloroplasts, or chromoplasts, originated from the radiant portion, the pigmented eye-spot from the attractive portion, and the flagella from the junction between the attractive and store portions. These flagella are the anterior, the forerunners of the nervous system, and the guiders of the products. Whilst the eye-spot is the antecedent of the eyes, and the controller of the flagella.

The first two forms of reproduction of which these cells are capable of undergoing, are binary and multiple fission. In both these forms the fission is mitotic; i.e. the nucleus chromosome splits into a number of chromosomes, each of which splits longitudinally, and the halves separate and move to opposite poles, where they associate together to form two daughter nuclei; when the fission is binary, and more when the fission is multiple.

The sparker-off of the meiosis is " activity's " description of the pulsatory cycles, in the cell, in the first half of which the store portion expands and the radiant portion contracts and the reverse in the second half. The " activity " and hydrogen and chemical compounds liberated from the store portion, as it expands, are accepted by the radiant portion as it contracts. Upon the radiant portion expanding " activity," hydrogen and chemical compounds are radiated to the store portion, which earlier had contracted and become unreactive. The radiation renders the portion reactive, and it is when this stage in the description of the pulsatory cycle is reached, that the secondary one of reproduction begins. In verification of this being the order in which the events occur, is the finding that reproduction occurs more commonly during the night than the day and the autumn and winter than the spring and summer. These are the times in the diurnal and animal pulsatory cycles, when the store portion is more contracted than expanded. When the portion is more expanded than contracted, which happens during the times the other secondary cycles are being more commonly described. And as the description of these other secondary cycles is more concerned with growth and reproduction, it will be seen how close the two are shown to be connected.

When a nucleus leaves the radiant portion, wherein it was born, to live in the store portion, it takes with it the properties of the radiant portion, and acquires those of the store portion on coming to live in the portion. In the splitting of the chromosomes and the moving of the halves to opposite poles, these properties run the risk of becoming disproportionately represented in the daughter nuclei; particularly in those which are formed in multiple fission. In this form of reproduction, some of the cells come to possess more of the store portion and others more of the radiant portion. Being opposite, and the properties of the store and radiant portions being those of femaleness and maleness respectively, the pairs conjugate, doubtless primarily for the purpose of establishing equilibrium.

In the binary and multiple fission types of reproduction, the process is referred to as being " asexual." Whilst the type in which two different kinds of cells conjugate, the process is referred to as being " sexual." The daughter nuclei in the first two types are called " zoospores " and in the third type, " zygospores."

When two products possessing opposite properties conjugate, whole chromosomes move to the opposite poles in the first division the mother cell, or zygote, undergoes; it being first in the second division that the process becomes and continues to be mitotic. This change in the way the products of conjugation divide, is called " meiosis "; and perhaps, its main significance is the effect the change has in increasing the disproportionateness of the properties possessed by the two different daughter nuclei. In the formation of the mother cell, or zygote, the two nuclei fuse to form what is called a " diploid " nucleus; for the reason that the chromosomes of each remain distinct. The nucleus remains diploid in the first division of meiosis, but in the second division, which is mitotic, the nucleus is restored to its original so-called " haploid," or single, state.

The factor determining whether the type of reproduction occurring in the flagellates be asexual or sexual, is the degree of contraction the store portion is made to undergo, in the description of the pulsatory cycle. The further it is made to contract, the more likely is the type to be asexual and the less sexual.

As evolution proceeds the daughter nuclei of the zygospores become more distinctive and specialised. The daughter nucleus, possessing more the properties of the store portion, becomes a single cell, receiving the name "oosphore". Whilst the one possessing more the properties of radiant portion, becomes a multiple of cells, receiving the name first, of "spermatozoids," in the vegetable products and of "spermatozoa," in the animal products. These more distinctive sexual cells are called "gametes" female and male respectively. And the body, formed by a single male gamete conjugating with the female gamete, is most commonly known as the "oospore." In all instances the gametes are haploid and the oospores diploid, which the latter remain until the meiosis they eventually undergo, results in the bodies produced being rendered haploid.

In this evolution of zygospores into female and male gametes, the attractive portion entered more into the production of the female gamete, in the animal than in the vegetable products. In the sexual reproduction the flagellates undergo, the attractive portion enters into the production of the zygospore, through whole cells being involved in the formation of the body. But as evolution proceeded, in the vegetable products, the attractive portion ceased to play a role in reproduction. This situation arose as a result of the store portion becoming increasingly more contracted as evolution progressed. The disappearance of the attractive portion through absorption, as the portion undergoes the same physico-chemical change in the description of the pulsatory cells as does the store portion, is not the only change which occurred, following the running parallel of the contraction of the store portion with the progress of evolution, as will be seen below.

The re-entry of the attractive portion, in animal reproduction is evidenced by the appearance of the benzene ring in the oestrogenic preparations; suggests the possible entry of the female gamete into the store portion, occurring indirectly from the radiant portion through the attractive portion. Should this be the route taken, an explanation would be afforded for the female sex exhibiting predominantly the function of attraction, in addition to the one of storing, as opposed to the male sex, which exhibits predominantly the function of radiation. The male gamete undergoes many more changes in its evolution than does the female gamete. This difference is to be accounted for, by the ever increasing contraction the store portion undergoes, in the vegetable products, rendering the changes occurring more uniform. And, through the effect this contraction has on causing the radiant portion to expand, increasing the changes occurring in the portion. In the next type of reproduction to be found in the vegetable products, the products themselves are found to exist in two states, the diploid and the haploid. The plant in the diploid state produces zoospores and came to be called the " sporophyte," whilst the one in the haploid state produces gametes, from which it derived the name " gametophyte." This type of reproduction arose, in the first place, from the increasing contraction the store portion undergoes in the evolution of the vegetable products. And, as the contraction increased with progress of evolution, the type, which is referred to as the " alternation of generations," became the characteristic one of the products. The factors responsible for this contraction were: first, the fixed position plants were obliged to take up, and one which necessitated their having to be dependent for their food on that that comes their way; and second, on having to leave the water, in which they were born, to come to live on land. The third factor, in being obliged to form oxygen from carbon dioxide, is quite subsidiary to the two just mentioned.

The more contracted the store portion becomes the further the gametophyte becomes supressed; the male more rapidly than the female. For the reason that what the female suffers is through absorption, whereas what the male suffers is through liquidation, which is caused by the over-expansion the radiant portion is made to undergo.

The progression of the absorption of the female gamete, or oospore, began in the Algae, results in its being transformed into the haploid embryo-sac in the seed plants. Whilst that of the liquidation of the male gametes, or spermatozoids, results, after their first having been rendered non-motile, in their becoming transformed into pollen-grains.

In the beginning of this type of reproduction, which made its first appearance in the seaweeds, both gametes are liberated from the products in which they are produced. When conjugation occurs, preference is shown for a male gamete from another plant, rather than from the same one, to fuse with the female gamete. So long as the contraction of the store portion does not exceed the normal, the oospore resulting from the conjugation develops into the plant, wherein the precursors of the gametes become transformed into diploid gametophytes. The fucoids are the seaweeds in which this life-cycle, resembling most closely the one that is most commonly described in the animal products, can best be studied.

Over-contraction of the store portion is avoided in these vegetable products by their having a freedom of movement, which enables them to obtain food those in a fixed position are unable to do.

In the other vegetable products, as the contraction of the store portion increases, and the asexual, or sporophytic, part of the plant grows more important than the sexual, or gametophytic, part, both gametes remain in the plant. The gametes may remain in the same plant, when the product is called " monoecious," or be found in different plants, when the product is called " dioecious." Only the male gametes are moved from the positions in which they are formed, as opposed to the female gamete whose position is never changed. The female gamete is a circular body, whereas the male gametes are composed of a head and tail, the latter of which is lost before the head in the liquidation the products are made to suffer. The head is oval, or round, in shape and the body in which the nucleus has its home. The tail is linear, or spiral, in shape and composed of fibrinogen, or fibrin, names which are given to the proteins found forming the basis of the junction between the store and radiant portions, from which this part of the product originated. The tail is attached to the posterior part of the head, which it propels; and, being opposite to the flagellum, which originated from the junction between the attractive and store portions, this part of the gamete is often referred to as the posterior flagellum or flagella, as similarly to the anterior there may be more than one.

Whilst on the subject of the spiral shape taken by the protein constituting the basis of the first junction, an explanation might be given for the spiral construction of the plant-apices, which apparently still remains a problem. It must be remembered that at the time these plant-apices are being constructed, the store portion is over-contracted, and needing reactivating, which can be effected solely by the radiant portion, on account of the absorption the over-contraction has caused the attractive portion to suffer. And this reactivation is effected by the passage of protein from the radiant portion through the junction to the store portion.

There still remain to be mentioned two other results of the varying degrees of over-contraction the store portion of the vegetable products undergoes. The first is the retrogression some of the Algae undergo to produce fungi, and their bacilli, sub-bacilli and cocci. In this ladder the gametophytic part of the plant undergoes infinitely greater depression than occurs in the other vegetable products. So much so that by the time the lowest rung is reached, the plants on it are found to contain no more than the nucleus that every cell possesses.

The second is the third type of reproduction to take place in the vegetable products, and to which the name " vegetative " has been given. This type is characterised by the ability plant-apices have to develop into similar vegetable products to those in which they were formed.

Although animals are able to move about in an oxygenated atmosphere, made for them by plants, and can search for the food they need, which is already prepared for them, it is not so much for these factors as such, that reproduction is simpler and more sexual than asexual in them than in the vegetable products. But to the influence these factors have in enabling the store portion to unite with the radiant and attractive portions, without the former having to suffer liquidation through being over-expanded, and the latter absorption through being over-contracted. Nuclei are the first organs to be involved in animal as in vegetable reproduction, and they may undergo the same changes; but in addition to these changes, there are others which are not to be observed occurring in plants. In one such instance single cells are to be found containing two nuclei, one large and the other small. The macronucleus stains badly, is non-motile and possesses the properties of the store portion. Whilst the micronucleus stains deeply, is motile and possesses the properties of the radiant portion. This small nucleus may have a flagellum attached to it, which it loses before approaching and fusing with the large nucleus. Following the fusion an asexual reproduction cycle is described in the product. In another instance two such cells may conjugate, when the first change to occur is degeneration of the large nuclei. This change is followed by two divisions each of the small nuclei. Of the four nuclei formed in each cell three degenerate, leaving the remaining one to divide again. Following this division one of each of the two nuclei formed migrates into the opposite cell, wherein it fuses with the nucleus which remains in the cell. The fusion results in the cells separating, and once this has occurred the fused nuclei; first divide several times; and second, each of the two cells divides twice to form four unicellular animal products. In this form of reproduction the small nucleus, remaining in the cell, acquires the properties of the store portion, whilst the other that leaves the cell to migrate, is the nucleus which retains the properties of the radiant portion. The behaviour of the two nuclei in the animal products, has only to be compared with that of the corresponding organs in the vegetable products, to show the dominance maleness has over femaleness in the former and the reverse in the latter. Further it confirms that this dominance held by femaleness in the vegetable products, arises from the way in which the store portion is made to over-contract and the radiant portion to over-expand.

The same difference is to be noted in the type of reproduction occurring in animals in which the organs involved are distinctive and specialised gametes instead of nuclei. The male gametes, in the forms of either Spirochaetae or Spermatozoa, are more active than their opposite numbers in the vegetable products. Indeed, there are precursors of Spirochaetae, male gametophytes, which enter endothelial cells in which to develop into still more active male gametes. While the female gametocyte leads an independent existence. Spermatozoa remain motile and they undergo no transformation into other bodies, as do the Spermatozoids in the vegetable products.

Despite the male gamete being more active in the animal than in the vegetable products, it remains still unable to reproduce by itself as the female gamete is able to do. Reproduction by the female gamete is called "parthenogenesis," the peculiarity of which is that only female and never male are the products of the process. This result shows definitely, that the female nucleus and gamete is the more important member of the two pairs of organs. From this difference the inference may be drawn, in the light of both the members having originated from the radiant portion, that the expansion and contraction the store portion undergoes in the description of the pulsatory cycles, exceed in importance the contraction and expansion the radiant portion undergoes in the description of the secondary cycles.

Having dealt with the effect the physico-chemical changes, namely with those the store portion of the protein undergoes in the description of the pulsatory cycles, have on the description of the reproductive cycles. This is the place, where it might be profitable to say something about the effect the changes produced in the latter cycle, have on the animal products themselves, particularly man, and their progeny, in health and disease. Should, in the description of the animal pulsatory cycles in the vegetable products, expansion of the store portion tend to exceed contraction in the spring and summer, more male plants will be produced than female plants; and, the opposite when the physico-chemical changes are reversed.

As protein evolves it becomes more condensed, which means that the functions of attraction and storing tend to be exhibited more predominantly than the function of radiation. Seeing that the exhibition of the former two functions is more the property of the female sex, whilst that of the latter is more the property of the male sex, the inherited tendency exists for more females than males to be born. Fortunately on this difference " climate," the factor responsible for the description of the pulsatory cycles, has a tremendous influence.

Protein has evolved further in man than in the other animals; and one of the results is the way in which the monthly pulsatory cycle has made itself so apparent in woman. When the contraction exceeds the expansion occuring in the month is the time the bleeding occurs. And why the physico-chemical change has this effect are: first, that both the female sex organs and the vascular system originated from the store portion; and second, that the dilation of the blood vessels in the organs results in blood permeating their walls. Should the combined proteins, at the time of conception, be more expanded than contracted, the child to be is more likely to be a boy than a girl; and, vice versa when the protein is more contracted than expanded.

In disease, should the protein in the mother be made to undergo abnormal physico-chemical changes early in pregnancy, the foetus suffers more than the mother; and, the other way round when the abnormal physico-chemical changes occur late in pregnancy. In the event of over-expansion being the principal abnormal physico-chemical change the protein is made to undergo in early pregnancy, the deformities more apt to arise are of an hypertrophic nature; such as the formation of supernumerary parts of the body. Whereas, when the abnormal physico-chemical change is the one of over-contraction the deformity is more likely to be of an atrophic nature; such as a loss of one or more parts of the body.

In the manifestations presented by the mother late in pregnancy, the nervous system is the organ most likely to be affected, owing to the protein being at this period least relatively over-expanded and most relatively over-contracted.

The greater tendency the protein shows in pregnancy to become more relatively over-contracted than over-expanded, affords the explanation for toxaemias in pregnancy occurring more commonly in women who are carrying female than male foetuses.

Should confirmation ever be able to be obtained of the above-mentioned present beliefs; that both kinds of nuclei and gametes originated primarily from the radiant portion; that the kind of nucleus and gamete which acquired, in addition, the properties of the store portion became the female; and that it was the acquisition of the double properties which enabled the female to develop into the complete whilst the male can do no more than reproduce its nuclear and gametal forms. These deductions must be drawn. First, that it was pure arrogance on man's part ever to have thought himself to be the superior of the two sexes; and, when fear drove him to look to a superior authority for guidance, to have pictured him in his own form. And second, that the most fundamental mistake man ever made, was in believing he could ever have been born of a virgin.

Until religion is brought up to date with science, man's attention will never be able to be focussed on its precepts, which have always been the more important and the more neglected of its contents.

## THE INFLAMMATION-CANCER CYCLE

The reproductive is not the only cycle for which the nucleus is responsible. The inflammation-cancer cycle is another and it came into being through the nucleus being the only organ, that not only became a constituent of every cell but also the most important one. The importance is revealed in the relationship which came to be established between the nucleus and the cytoplasm; especially, with the store portion of the latter. Indeed, it was a result of the establishment of this relationship, that this cycle became responsible for one of the divisions of disease being into the two morbid conditions of inflammation and cancer. The other two divisions are the stages and the manifestations, about which no more need be said than is relevant to the sub-matter in hand, as both have been dealt with fully in my other writings. The relationship between the nucleus and the cytoplasm of cells became established, as a result of a concatenation of events, which occurred in somewhat like the following sequence. In the first place, the nucleus was the first organ to appear; presumably after the radiant portion of the protein had begun to take the lead in the secondary cycles being described in the products; and the results of which had already begun to be influenced by the manner in which the pulsatory cycles had earlier been described. In the second place, the nucleus originated from the radiant portion, wherein it remained the sole organ capable of reproducing itself; and of migrating into the store portion, wherein the organ acquired the properties of the portion. In the third place, this possession of opposite properties by the same organ, not only brought the two kinds of nuclei, but also the store and radiant portions into closer apposition, than could otherwise have been possible. Finally, the closer relationship that thus came to be established, first between the two portions, and later between the two kinds of the same organ, made the junction between the store and radiant portions not only the most important of the three in protein, but also the one that cannot be by-passed in the cycles which are described in the products. Indeed, the abnormal physico-chemical changes this most important area in protein, and the organs which originated from it, are made to undergo in disease, comprise those which are responsible for what have come to be called " inflammation " and " cancer."

## INFLAMMATION

The abnormal physico-chemical changes the junction between the store and radiant portions is made to undergo in the three stages of disease, are responsible for three distinctive sets of results. Catalogued, these results include the formation of the so-called " viruses," and the accumulation of red blood corpuscles and polymorpho-nuclear leucocytes at the seat of inflammation, in the acute stage. Clotting *in vivo* and the accumulation of mononuclear leucocytes at the seat of inflammation, in the sub-acute stage. And the reproduction of the nucleus and the cell in which it resides to form innocent new growths, in the chronic stage.

The so-called " viruses " are no more than the products of the over-contraction the radiant portion is made to undergo in the acute, or seasonal, manifestations of disease. The junction between the store and radiant portions is the area in the protein to be most affected in the over-contraction. As the over-contraction follows the over-expansion " climate " has earlier caused the store portion to undergo, the variation to be found in the over-contracted products can be attributed primarily to the degree of over-expansion " climate " makes the store portion undergo. As nothing is more variable than " climate," and it is variability that excludes any two endemics, epidemics or pandemics being exactly similar, it follows that no limit will ever be able to be placed upon the number of so-called " viruses " capable of being produced. To add to the difficulties arising from these variabilities, are the additional ones of the grade of the condensation reached by the protein, and of the season and time of the year and the time of the day, and of method which is employed to obtain the product.

Products allied to so-called " viruses " are the inclusion bodies.

Inclusion bodies are to be found most commonly in the cytoplasm of the epithelial cells, in smallpox; and in the cytoplasm and nucleus in the nerve cells, of so-called " Borna disease " in horses. The explanations for these selective findings, lie solely in the different abnormal physico-chemical changes the store portion of the protein is made to undergo in the production of the products. For the skin to be the organ in which the acute, or seasonal, manifestation of disease is made to appear, the store portion needs to be more over-expanded and to become more relatively over-contracted, than is required to make the manifestation appear in the portal and respiratory systems. And to be still more over-expanded and to become still more relatively over-contracted, than is required to make the manifestation to appear in the nervous system. For the reason, that the cyclical manner in which the organs originated from the reactive portions of the protein, were the pulmonary and portal systems from the radiant portion, and the cutaneous and nervous from the attractive portion.

The horse is the animal in which the store portion of the protein is normally most expanded and relatively contracted.

The explanation for the erythrocytes and the neutrophils being the corpuscles selected to collect in the area of acute inflammation, resides in their having been the first of their kind, not only to have originated from the junction between the store and radiant portions. But also, in consequence, from the part of the junction which remains over-expanded instead of becoming over-contracted, which is the abnormal physico-chemical change the lower part undergoes in response to the over-expansion the store portion has earlier been made to undergo. Expansion of an area in the protein has a growth effect on the organ which originated from it; whereas the action of contraction is the opposite, the suspension of the process, which in disease can become translated into hypertrophic and atrophic respectively.

## THE SUBACUTE STAGE

In the subacute stage of disease, over-contraction is the most characteristic abnormal physico-chemical change the store portion is made to undergo. The change follows the one of over-expansion the protein is made to undergo in the acute stage, and it is as variable as the latter, and it is followed by over-expansion of the radiant portion. When all three abnormal physico-chemical changes reach much the same level, they are made to do following the addition of an anticoagulant to plasma to prevent it from clotting, a somewhat similar dark-ground picture of the protein particles may be obtained. The particles are diminished in number and mobility, and increased in size and refractility; and they may be surrounded by a ring, to which are attached threads in either a colloid or a crystalline state. Should, in addition to the kations released from the radiant portion when it is made to over-expand, sufficient carbon dioxide collect in the blood to send these abnormal protein particles into solution, their ultimate return to the colloid state causes them to clot *in vivo*. The clotting occurs in the veins most often, first in the left posterior tibial, and the pulmonary arteries, owing to these being the sites in which the kationic-acting carbon dioxide is most likely to collect. On the return to the colloid state in vivo, following the addition of a more powerful kation than carbon dioxide, the protein particles reappear; first, in the form of fibrin threads, which become rapidly covered by myriads of small non-refractile and non-motile particles. Second, these particles become enclosed in a fibrin network, which, on contracting, squeezes the particles through into the fluid around, wherein they regain the forms the particles had before the addition of the anticoagulant. In the veins and the pulmonary arteries the changes the particles undergo, do not progress beyond those mentioned above as occurring in the first stage. The changes the protein particles undergo *in vivo,* following, first the use of an anticoagulant, and second, of a kation, are repeated in the bacteriophage formation in bacteria.

The factors taking the place and having the action of the anticoagulant, include the differences between the conditions under which the bacterium was living at the time of its capture, and those under which it was subsequently made to live; and whether, following cultivation, the product was filtered or not. The more the conditions under which the micro-organism was living, irrespective of those under which it was subsequently made to live, the more readily is its store portion made to over-expand and over-contract. And the effect filtration has is to increase the susceptibility of the portion to undergo these abnormal physico-chemical changes. The bacterium I found to react most readily to these conditions was the *Pseudomonas pyocyanea.*

This micro-organism was the first true bacterium to regress from the fungus *Rhizobium leguminosarum,* as the *Bacillus faecalis* alkaligenes, which preceded it, continued to retain more the properties of its fungal parent than to acquire those of the bacterial products. Furthermore, the *Pseudomonas pyocyanea* was the bacterium which appeared to inherit a greater likeness for the water, in which its antecedents the *Algae* had been born, than any other. But the water needs to be fresh, and not to have suffered stagnation, or filtration. The fondness of the micro-organism for water, is evidenced by its being one of the commonest to be found in the excreta of animals, which live the whole, or part, of their time in this medium. And by the frequency with which the bacterium is found to be an invader in the wet, or moist, areas of inflammation in disease, affecting plants, animals and man. All filtered, and thought to be sterile, water, should always be looked upon with suspicion, as it is apt to contain bacteriophages, which develop rapidly into the *Pseudomonas pyocyanea,* on reaching a favourable medium.

This description of the way in which bacteriophages are alleged to be formed, lends strong support to the belief that fungi regressed from *Algae,* and bacilli, sub-bacilli and cocci from fungi. And that the regression was caused by desiccation over-expanding and over-contracting the store portion of the *Algae,* the latter abnormal physico-chemical change of which occasioned the radiant portion to over-expand, in the process of which the chloroplasts were liquidated. The important role over-expansion of the radiant portion plays in the production of bacteriophages, is emphasised still further by the finding that all the nucleic acid in the products is of the deoxyribose variety.

The first part of the junction to over-expand in the subacute stage of disease, is the one from which the mononuclear leucocytes originated. This area is situated below the one from which the red blood corpuscles and the polymorpho-nuclear leucocytes originated, and it remains over-contracted in the subacute stage. This difference accounts for the replacements in the areas of inflammation, of the occupants of the second group by those of the first.

The first mononuclear leucocytes to put in an appearance are the lymphocytes. As the store portion grows relatively more over-contracted the large mononuclear leucocytes, or plasma cells, take the place of the smaller variety. Then appear the endothelial cells, the parent cells of these white blood corpuscles; and, the longer the store portion remains relatively over-contracted, the greater grows the inclination of the endothelial cells to coalesce to form giant cells.

The action of the neutrophile leucocytes is similar to that of the water-soluble vitamins, some of which their cytoplasm contains; and whose hydrogen atoms, on being released, place a check on the store portion's over-expansion, and correct the over-contraction, should the portion have begun to undergo this abnormal physico-chemical change. The action of the mononuclear leucocytes, whose principal constituents of their cytoplasm are the amino acid complexes, is to expand the over-contracted store portion of the protein in the blood, in which the host's main resistance lies.

## THE CHRONIC STAGE

In the chronic stage of disease the store portion is both over-expanded and relatively over-expanded. On other words, these are the same abnormal physico-chemical changes, as those the portion undergoes in the acute stage. But in reality there exists a great difference between the two and, the test which shows it most clearly is the sedimentation rate of the red blood corpuscles. In the acute stage of disease the rate is increased, whereas in the chronic it is either unaltered, or diminished. From the rate remaining more or less unchanged in the chronic stage, the deduction may be drawn, that the over-expansion which the store portion undergoes is unreactive; in which state it fails to be reflected onto the radiant portion to cause it to over-contract. When the change is rendered reactive it is apt to be reflected directly on to the nucleus; first, because the cytoplasm is the part of the cell to bear the brunt of the attack in the subacute and acute stages. And second, because the nucleus was the first organ to originate from the radiant portion. This stimulation of the nucleus causes it, and the cytoplasm of the cell in which it is lodged, to multiply to form an innocent new growth. The organ in which most of these new growths are formed is the skin. For the reason, that the extent to which the store portion is over-expanded in the chronic stage of disease, approaches most closely to the level reached in the acute stage, which causes the cutaneous to be the type of influenza produced. And in the most severe form (small-pox), inclusion bodies to form in the cytoplasm of the epithelial cells.

The cutaneous new growths may be composed of the epithelial cells, or the hair follicles, sebaceous glands and sweat glands, singly, or combined. No sites exist, in which the latter are more commonly to be found, than in the orbito-facial and naso-facial folds. Most of these facial new growths are atavistic of the specialised glandular structures to be found in so many of the animals, which evolved before man.

## THE BRIDGE BETWEEN INFLAMMATION AND CANCER

This bridge begins to be crossed when the rendering reactive, the over-expansion the store portion undergoes in the chronic stage, extends to the over-contraction the portion subsequently undergoes. So long as the over-expansion of the store portion remains in the ascendancy, not only does the nucleus of the cells in the organ affected continue to be stimulated directly but also, when the peak is reached, the organs found to be those most frequently affected, are the mononuclear leucocytes. The explanation for this selective action, is that the junction between the store and radiant portions, from which the nucleus originated, is similarly the area in the protein from which the mononuclear leucocytes had their origin.

The moment the over-expansion of the store portion became automatically replaced by over-contraction, the nucleus and the cytoplasm, of the cells in the organ attacked, began to degenerate. The manifestations of disease for which these abnormal physico-chemical changes may be held responsible, are the aleucaemic and leucaemic lymphocytomata. The aleucaemic manifestations are more inflammatory in nature and the leucaemic more cancerous. No sharp line of demarcation can be drawn between the two, as a manifestation beginning as an aleucaemic one may end by becoming leucaemic. A possible explanation for the change is the stimulating action exercised either by an extension of the over-expansion of the store portion or by the radiant portion on first becoming over-expanded in response to the store portion's over-contraction, or both.

The inflammatory manifestations with which those under consideration are closely connected, include the various forms of Mycosis fungoids, in man and so-called " white-eye " and "paralysis" in fowls. The main feature about these manifestations, is the comfortable way in which they fit into the cycle being described. The skin is the organ most commonly attacked by the forms of Mycosis fungoides, the eyes by so-called "white-eye," and the central nervous system by so-called "fowl paralysis." The order is the one in which the organs originated from the attractive portion, travelling from the radiant to the store portion. The organs are those attacked in the acute stage of disease, when the attractive portion is made to over-contract. The organs are those in which inclusion bodies may be formed, and last to be attacked before the blow falls onto the nucleus.

## CANCER

The most obvious changes to be noticed to take place in the nucleus, when a new growth becomes malignant, are those affecting the nucleolus. This part of the nucleus contains both nuclear and cytoplasmic material. As the nucleolus comes to be the part of the nucleus, upon which the attack is centered, it becomes more prominent; more independent and over-contracted. The over-contraction is judged by the nuclear material staining more deeply, and the cytoplasmic material being more refractile, and by the subsequent changes the nucleolus undergoes being those of division and subdivision.

The nucleolus, on reaching these states in its development, behaves like a pathogenic micro-organism by becoming parasitic on its host. The parasitism is evidenced by the nucleolus entering the circulation, and settling down in the organs of its choice, to form metastases. This likening of the nucleolus to pathogenic micro-organisms, does not end in the formation of metastases, because, when examined further, the ways in which metastatic and simple cancer are produced, are seen to resemble the two forms of regression the vegetable products underwent. Metastatic cancer resembles the regression the *Algae* underwent into fungi, and the latter into bacilli, sub-bacilli and cocci; and, simple cancer that the gametes suffered, the male more than the female, the *Algae* underwent which developed into seed plants.

In metastatic cancer, the protein in the blood undergoes the same abnormal physico-chemical changes as it does in the subacute stage of disease, in which one or more of the micro-organisms normally residing in the large intestine, become pathogenic. No more common sites exist, in which simple cancer is to be found, than in the prostate gland and the uterus; and here again, more commonly in the former than in the latter.

Why cancer generally should appear at different periods to attack different organs, would seem to be governed by the same factor, as the one which decided the organ to be attacked by the acute, or seasonal, manifestations of disease. This factor is the extent to which the store portion of the protein in the blood is over-expanded. Finally, as the changes the nucleus undergoes in the chronic stage and cancer, are a repetition of those the protein undergoes in the acute and subacute stages of inflammation, no surprise should be occasioned by finding so-called " viruses " in the nucleus in leucaemic and cancer more especially, as these bodies are most commonly formed in the site in the radiant portion from which the nucleus originated. And the finding of cancer cells coagulating, once the condition has become metastatic, much as the protein particles in the blood are prone to do in the corresponding stage of rheumatoid arthritis.

## THE CHEMOTHERAPEUTIC CYCLE

My experience leads me to believe, that the aim of chemotherapy is to restore " activity " to its normal path in the cycles, which the same agent has caused to be abnormally described in disease, in the protein in the sap and blood. The protein, primarily, in the cytoplasm, and secondarily, in the sap and blood, was the first product in the plants and animals in which all the above-mentioned cycles came to be described. And further, it was the way in which these cycles were described, that enabled organs to originate from the three portions of the cytoplasm and protein. This origin is confirmed; first, by the basic chemical constituents of the organs being the same as those of the portion of the cytoplasm and protein from which they originated; and second, the organ attacked in disease is determined by the kind of abnormal way in which the cycles are described in the cytoplasm and protein.

The radiant having been the first portion from which organs originated, is confirmatory of the lead in the metabolism this portion gained over the store and attractive portions. Similarly, the abnormal physico-chemical changes the store portion is made to undergo in disease, are those which determine the results of this lead, and the organ in which the manifestation appears.

The leadership of the radiant portion, in the description of the secondary cycles, was gained by the amino acids, which had developed from the organic acids in the store portion, forming complexes and giving rise to other chemical compounds, the first of which became the leaders and the second their assistants. The complexes were formed by the amino acids combining in differing sequences, to produce the polypeptides. The sequence rather than the combination is the factor, which made these complexes the leaders; and each to become associated more closely with its own particular assistant.

The actions of the leaders would appear to be; first, to maintain the areas in the protein which they control in their normal physical state; and second, to give directions to their assistants. These directions consist in donating atoms and chemical compounds to the store and attractive portions, as the radiant portion expands; and accepting these atoms and compounds as the portion contracts. As the physico-chemical changes the radiant portion undergoes in the description of the secondary cycles, are influenced by the way in which the pulsatory cycles are described, it will be seen how it happened that the results of the leadership gained by the radiant portion came to be governed by another factor. This other factor is the physico-chemical changes which the store portion is made to undergo in the description of the pulsatory cycles.

The reasons for the store being the portion to be most affected in the description of these cycles are; that it was the first portion to appear; the one from which the radiant portion developed; and the junction between the attractive and store portions being the area in the protein where the cycle commences to be described. In the description of these pulsatory cycles, the store and attractive portions expand and contract whilst the radiant portion contracts and expands. The acceptance of atoms and chemical compounds, released from the store and attractive portions as they expand, by the radiant portion as it contracts, would never ordinarily be interfered with but for one factor. This factor is that few consecutive pulsatory cycles are described in exactly the same way. The different ways arise from the ever varying quality and quantity of the " activity " which, after penetrating the junction between the attractive and store portions, describes the cycles. On one occasion more " activity " is released from its storage depot in the junction than on another. And, it is on the quality and the quantity released, that the extent to which particularly the store portion, is made to expand and contract. In the event of the store portion being made to over-expand and over-contract, the radiant portion over-contracts and over-expands. These abnormal physical changes the radiant portion is made to undergo, naturally affect the way in which the atoms and chemical compounds are accepted and donated.

In the physico-chemical changes the protein undergoes, in the description of the primary and secondary cycles, the physical changes the store and attractive portions undergo exceed in importance the chemical changes; whereas in the case of the radiant portion the positions are reversed.

Forming the basis of evolution, or of the progress and regress, the protein undergoes, as do the grades of expansion and contraction to which its store portion is subjected in the description of the pulsatory cycles, accounts—not only for any specificity that may be surmised to exist between the leaders and their assistants in the radiant portion, and the changes occurring in the store and attractive portions in the description of the secondary cycles; but also between the grades of expansion and contraction the store portion undergoes; and, the micro-organism which becomes parasitic in the subacute stage of disease. These micro-organisms include the fungi, bacilli, sub-bacilli and cocci, which were the products of the regression the unicellular plants underwent; and, the *Protozoa,* which were the products of the regression the unicellular animals undersent.

Hydrogen is the first atom and phosphoric acid ($H_3PO_4$) the first compound to be released from the store portion as it expands. The hydrogen is accepted most commonly by a water-soluble vitamin and the phosphoric acid by a nucleotide as the radiant portion contracts.

The assistants in the radiant portion are frequently combinations of water-soluble vitamins and nucleotides. The sequence in which the amino acids are arranged in the polypeptide, which is the leader in this acceptance; and, the donation of the atom and chemical compound when the radiant portion comes to expand, is apparently unknown. The composition of many of the assistants is known, but the same cannot be said of the leaders; and, what is known of both shows their action to be a group rather than a singular one. The leaders whose composition is best known, are curiously enough, those which form the basis of the integrated endocrine glandular system, which originated from the radiant and attractive portions. As the system evolved from the organic part of the sense-organ mechanism, which is divisible into specialised and generalised parts, and as the ductless glands which evolved from the specialised part, are those which originated from the three junctions—the most important areas in the protein,— it is possible, that there are not as many groups of leaders as there are of assistants. Supporting this possibility, is the regression the integrated endocrine glandular system has suffered since its evolution from the organic part of the sense-organ mechanism.

In disease, the first demonstrable physico-chemical signs of which are the over-expansion the store portion undergoes, and the extra " activity " hydrogen and phosphoric acid that come to be released from the portion. Should the extra " activity " released become ionised in the body, fever is produced. In the over-contraction the radiant portion undergoes, this extra hydrogen and phosphoric acid are accepted, but, instead of being available for redonation to the store portion, when the portion over-expands, they are kept stored and rendered unreactive in the over-contracted part of the portion.

These abnormal physico-chemical changes which the store and radiant portions undergo, are those which are responsible for the acute stage of disease. Owing to " activity " being the principal cause of the over-expansion and over-contraction the store and radiant portions undergo, and to the pulsatory cycles being divisible into diurnal, menstrual, annual, biennial and multiennial, the manifestations produced are referred to as being seasonal. The manifestations include mosaic in plants, distemper in animals and influenza in man. The organ in which the manifestation appears in animals and man, is determined by the extent to which the store portion is relatively over-expanded. The more relatively over-expanded is the portion, the more likely is the pulmonary to be the system attacked; and, the less relatively over-expanded the nervous system, the portal and cutaneous coming in between. In due course, the store portion becomes over-contracted and the radiant portion over-expanded. On the store portion becoming over-contracted, one or more of the micro-organisms resident in the intestinal tract become pathogenic. Those selected to become pathogenic are decided by the extent to which the store portion is over-contracted, and has been made to undergo this abnormal physico-chemical change, ever since the micro-organisms made the intestinal tract their home. The relationship between the over-contraction the store portion undergoes in the subacute stage of disease and the micro-organism which is rendered most pathogenic by the change has long become an inherited one. When the store portion becomes relatively over-contracted, rheumatism is the manifestation most commonly to be produced, and the organs to be most affected are the tissues and the vascular system. For the reason, that the store is the portion from which these organs originated. Once the store portion becomes more over-contracted than over-expanded, it remains longer in this abnormal physico-chemical state than does the radiant portion in the acute stage of disease. Indeed, the change may persist, particularly in the area most affected, even beyond the time when the greater part of the portion becomes over-expanded and relatively over-expanded, or undergo the changes which characterise the chronic stage of disease. The persistence of the over-contraction of the store portion into the chronic stage of disease, accounts for the degenerative changes to be noticed in the organs which originated from the portion. Whilst the later hypertrophic changes to be found in the organs, are the results of the expansive changes the portion undergoes in the chronic stage.

The fundamental differences between the three stages are. First, that the junction between the store and radiant portions is the area in the protein in the blood to be most affected in the acute and subacute stages; and the nucleus, which originated from this area, in the chronic stage. Second, that the organs which originated from the radiant and attractive portions, are affected more often than those which originated from the store portion in the acute stage, and the reverse in the subacute stage. Third, that the radiant portion of the cytoplasm of epithelial cells and of the nucleus, in addition, of nerve cells, in the acute cutaneous and nervous manifestations, may over-contract to form inclusion bodies; as does the junction between the store and radiant portions of the protein in the blood in the acute stage to form viruses. Fourth, that the over-expansion the junction between the store and radiant portions undergoes in the subacute stage, may result in the protein particles clotting *in vivo* and in phages forming in bacteria. Fifth, that so long as the changes the store portion undergoes in the chronic stage remain innocently expansive, the nucleus is stimulated to form new growths. But, that when they become malignantly expansive, viruses may make their appearance in the organ. When these malignant changes are superceded by corresponding contractile changes, degenerative changes occur in the nucleolus, causing it to behave as a parasite, and even to undergo the same changes as does a bacterium in phage formation.

## ACUTE STAGE

The way in which " activity " is deflected from its normal path in the description of the cycles in the protein; and, which seems to be most clearly revealed in the inability of the radiant portion, on expanding, to donate hydrogen and phosphoric acid to the store portion, suggests to me, that if ever a chemotherapeutic preparation be synthesised able to counteract the expansive action " activity " is exerting on the store portion, it will need to be a nucleotide, or polypeptide, resembling as closely as possible, those being rendered unreactive in the over-contracted radiant portion. If there is anything in this suggestion, greater support could even be produced to show how secure is the leadership the radiant portion holds in protein metabolism.

The nearest approach that has so far been made to placing a check on " activity's " over-expanding the store portion, is in the use of antipyretics. These preparations contain a benzene which, carrying so-called " positively charged " active groups, mainly methyl or hydroxyl, whose action it is to replace the hydrogen being released from the store portion; and, to reduce the oxidation the portion is being made to undergo as a consequence. The hydrogen released from the store portion can be replaced only by that that has been reduced in the radiant portion, and donated to the store portion as the radiant portion expands.

As the radiant portion fails to expand until the store portion automatically begins to contract, and as the latter is the signal for one or more of the micro-organisms resident in the intestinal tract to become pathogenic, an explanation is afforded for the sulphanilamide compounds and the antibiotics having a greater antibacterial than an antiviral action. In neither case is their action direct, but indirect through being able to donate to the store portion the hydrogen of which it had been deprived, whilst being compelled by " activity " to over-expand.

Why some antibiotics appear to have a greater therapeutic action when certain micro-organisms become more pathogenic than others, is to be explained as follows. In the first place, antibiotics are the natural components of the radiant portion of the animate products, in which they act as assistants to the leadership the portion exercises over the store and attractive portions. And in the second place, the assistantships came into operation, at different stages, in the regressive path the *Algae* took to become bacilli. In other words, a certain specificity exists between the period during which a certain derivative of a polypeptide assumed the role of an assistant, and that during which the regression the *Algae* underwent, resulted in the production of a certain bacillus. This relationship which exists between the development of the chemical compounds, constituting the basis of the radiant portion, and the production of the fungi, bacilli, subbacilli and cocci into which the *Algae* regressed, explains how the indiscriminate use of antibiotics has resulted in rendering certain micro-organisms resistant, others more pathogenic and fungi more parasitic. The closer the chemotherapeutic preparations in future approximate to the chemical compounds, constituting the basis of the protein in the blood, the greater will grow the potential harmful effect they are likely to have on the body, should they be injudiciously prescribed. In other words, the closer treatment comes to consist of treating like with like, the greater will become its toxic potential.

## SUBACUTE STAGE

Long before relative over-contraction of the store portion becomes the most characteristic abnormal physico-chemical change the protein is made to undergo, in this stage, not only is the change itself but also the accompanying one, namely, the over-expansion the radiant portion is made to undergo, responsive to the use of the antibiotics; particularly in the compounds most active group is a carboxyl one. This group has a contractile effect on the radiant portion; it is capable of donating hydrogen to the store portion; and, it counteracts the oxidation this portion undergoes when it over-contracts. But once the over-contraction has reached a certain level, the only chemotherapeutic preparations, at present available, having the power to correct the change, are those similar to and which exist in a more oxidised state, than those being most affected in the protein. The compounds to be most involved in the over-contraction the store portion undergoes, are the steroids. The same is true of the compounds best able to effect the needed expansion; the main difference between the two being, that the expansive compounds require to be more oxidised than those which have been involved in the over-contraction. Hence the use of the cortisone preparations, in the subacute stage of disease, and the greater efficiency of those which contain a fluorine atom.

The way in which the cortisone preparations act in the subacute stage of disease, resembles that of the immunological preparations, which expand the over-contracted globulin protein particles.

A rule may be made; that to contract over-expanded protein, dissimilar compounds, carrying opposite functioning active groups are required, whereas to expand over-contracted protein, similar compounds carrying slightly dissimilar active groups are needed.

The action of the higher metals in the porphyrins, is fundamentally similar to that of hydrogen in the water-soluble vitamins. And any specific action a trace element may be surmised to have, may be explained by the abnormal physico-chemical change it is able to correct, having made its appearance during the period in which the element became the active atom in the porphyrin carrying it.

## CHRONIC STAGE

Although over-expansion and relative over-expansion of the store portion, are the most characteristic abnormal physico-chemical changes the protein undergoes in this stage, nevertheless, the persistence of the over-contraction from the subacute stage, is the change which continues to stand most in need of chemo-therapeutic correction. The correction is undertaken either with preparations which act directly on the store portion, or indirectly by first expanding the over-contracted attractive portion. The purposes for which these preparations are most frequently prescribed, are to stimulate the cardio-vascular system; to increase the output of urine from the urinary system; and to prevent clotting from occurring *in vivo*. Some of the most active cardiac stimulants, having a direct action on the store portion, are steroids; and the way in which they expand the over-contracted portion, resembles that of the cortisone preparations. The resemblance between these glycosides and corticosteroids is further emphasised, by the tissues and the vascular system having been the first two organs to have originated from the store portion; and, by the frequency with which the vascular system is attacked in rheumatism. The cardio-vascular stimulants which expand the over-contracted attractive portion, include the common adrenergic compounds and the sympathomimetic drugs, whose vehicle is composed of a nitrite group ($NO_2$).

The direct acting diuretics are the mercurial preparations, of which mersalyl is a good example. Whilst those which act indirectly do not contain this metal, but instead sulphur and a nitrite group. The latter are commonly known as saluric acid compounds. The urinary tract, along with the genital system, e.g. the urogenital system, was the third organ to originate from the store portion. The explanation for the close association the cardiac stimulants and the diuretics show to exist between the store and attractive portions, lies in the physico-chemical changes these two portions undergo, in health and disease, being similar and the opposite to those the radiant portion undergoes.

Practically speaking, the drug best able to prevent the protein particles clotting in the veins and pulmonary arteries, is one which is composed of an aromatic vehicle carrying one or more active carbonyl groups. For these groups antidote better than any others the contractile-acting aldehyde group, which is the group in aldosterone, the steroid compound which sparks off the process. The area in the store portion which a hyperactivity of aldosterone causes to over-contract, is the radiant end, where begins the junction between the two portions. But, once the store end of this junction becomes over-contracted, no carbonyl groups are able to exert their expansive action upon it, for the two following reasons. The first, is the coming into play of the contractile-acting polypeptide active principle of the parathyroid glands, to which further reference is made below. And the second, is that the area in the store portion which the carbonyl groups are best suited to expand, is that from which the vascular system originated. Hence the explanation for the readiness with which haemorrhage may follow the injudicious use of anticoagulants.

The organs which originated from the attractive portion are not spared, when the over-contraction of the store portion involves the attractive portion. The organ to suffer most is the nervous system; and particularly so, when the organ, which is most affected by the store portion being over-contracted, is the vascular system. One of the most common nervous manifestations to be produced by the attractive portion being over-contracted, is Paralysis agitans. Methyl groups are depended upon to expand the over-contracted portion. But, owing to the frequency with which this group is only the end link of a larger group, which constitutes the basis of both the radiant and attractive portions, and is carried by an aromatic vehicle, great care needs to be exercised in the use of what are called anti-convulsants.

Although the primary action of these remedies is to expand the over-contracted attractive portion, their secondary action is to over-contract either this, or the radiant portion, or both.

This association, between the two reactive portions, is brought about by the independence which was gained by the expansive part of the attractive portion, the result of which was to make the contractile part of both reactive portions indivisible. This indivisibility explains, why so many of the anticonvulsants are described as having both a parasympathetic and a sympathetic action. And why, when they caused over-contraction, they can produce analgesia, anaesthesia and a cutaneous eruption (attractive portion) or vomiting (radiant portion).

When the over-expansion the store and attractive portions undergo in the chronic stage takes the place of the over-contraction, hypertrophic changes begin to make their appearance in the organs, which originated from the store portion. And psychoses are produced by somewhat similar changes occurring in the central nervous system, which originated from the attractive portion. Hypertrophic changes are found affecting the bones, first and most frequently the lumbar vertebrae, the uterus and prostate gland. So far no chemotherapeutic remedies are available for halting the process; but several are available for alleviating the signs and symptoms of the psychoses. One of the most active groups in the aromatic compounds, used for the latter purpose, is the contractile-acting peptide group (—CO—NH—). But, the compounds need to be used with caution, for fear of over-contracting one or other of the two reactive portions. Ultimate contraction of the over-expanded brain, is the action electric shock therapy has in involutional melancholia. Once the over-expansion and relative over-expansion the store portion undergoes in this stage comes to be reflected on to the nucleus, the first organ to be affected is most commonly the skin. First, because the attractive portion undergoes the same abnormal physico-chemical changes; and second, because the skin was the first organ to originate from the attractive portion. Upon the store portion becoming automatically over-contracted, the organ selected for attack is the mononuclear leucocyte, owing to its being the first organ whose nucleus contains a nucleolus, which originated from the radiant portion. Should the affected mononuclear leucocytes choose another organ, in which to carry on the leucaemic process, the skin, again, is the one most commonly chosen, and the manifestation produced, one of the forms of Mycosis fungoides.

On the store portion becoming relatively over-contracted, the attack is centred on the nucleolus of the nucleus. And the organ in which cancer appears, is influenced by the extent to which the store portion was earlier relatively over-expanded. It is possible that better derivatives of the leader in the radiant portion, which is involved in the over-expansion and relative over-expansion the store portion is made to undergo, may be found possessing a better therapeutic action than those at present in use. Little hope remains of being able to correct the abnormal physico-chemical changes, short of discovering the polypeptide which acts as the leader. Once the automatic over-contraction of the store portion gains the ascendancy, the employment of the corticosteroids have no more than a transient action; and still less hope exists of finding a chemotherapeutic preparation able to contract to over-expanded radiant portion, as none exercising this action sufficiently exist at present, as far as I am aware.

## THE CHEMOTHERAPEUTIC REMEDIES TO BE DERIVED FROM THE INTEGRATED ENDOCRINE GLANDULAR SYSTEM

The integrated endocrine glandular system is the representative in the vertebrates, of the organic part of the sense-organ mechanism in the invertebrates. The organic part of the sense-organ mechanism is divided into specialised and generalised sense-organs. The specialised sense-organs, and the ductless glands into which they developed, originated from the reactive part of the junctions between the three portions of the protein. Whilst the generalised sense-organs, and the ductless glands into which they developed, originated from the two reactive portions. No sense-organs, or ductless glands, originated from the unreactive, or store, portion. All the ductless glands contain one or more specific polypeptides, which act as leaders, and several active principals, which act as assistants. The specific polypeptides in the ductless glands, which originated from the three junctions, have as a basis calco-phospholipid complexes, whereas those, which originated from the radiant and attractive portions, do not possess these chemical compounds.

Comparing the action of the polypeptides in the ductless glands, which originated from the radiant portion, with those, which originated from the attractive portion, the former have both a contractile and an expansive action, especially those which originated from the first and second junctions. Whilst the latter have either the single action of expansion or the one of contraction. The radiant assistants have the single action of contraction, whilst the attractive assistants have also the single action of either expansion or contraction.

## PARATHYROID GLANDS

These glands originated from the reactive part of the first junction, which lies wholly in the radiant portion. Little is known of the chemical composition of either the specific polypeptides or the active principals. What is known, is that the extracts have a contractile action. This contractile action is exhibited normally, presumably by the hydrogen atom of the carboxyl group of the glands specific polypeptide, when " activity " is led from the expanding store portion into the contracting radiant portion. And under abnormal conditions, by the release of the calcium from the calco-phospholipid, which occurs when over-contraction of the store portion causes the radiant portion to over-expand. The result of this action of calcium is to cause clotting to occur. The timely use of an oxidising agent suffices to stop this contractile action of both the hydrogen and calcium atoms from being exhibited.

## THYMUS GLAND AND ISLET CELLS OF LANGERHANS

These ductless glands originated from the radiant portion; the former from the upper and lower parts, and the latter from between the two, as far as can be determined at present. The upper limb of the thymus appears to be connected with the metabolism of the mononuclear leucocytes, and the lower with the description of the urea cycle. The islet cells of Langerhans are connected with the description of the carbohydrate cycle. The extracts made from the glands have both a contractile and expansive action on the radiant portion, in which the former is much more marked than the latter. Whether the dual action is due to the same polypeptides in different chemical forms, or to different polypeptides, has yet to be determined. The sequence in which the amino acids are arranged in the thymus polypeptides is at present unknown. But from the little experience I have had of the use of the extracts in the treatment of certain manifestations of disease, I feel sure its discovery would well repay investigation.

The sequence in which the amino acids are arranged in insulin has been discovered. The main feature of the discovery is the finding, that the polypeptide is composed of two chains which are linked by a disulphide group. For it shows how intimately connected are the contractile parts of the radiant and attractive portions.

## THYROID GLAND

This gland originated from the radiant end of the second junction. The extract has an expansive action. This expansive action is undoubtedly due to the presence of the iodine atoms in the compounds. Although thyroxine has been recognised as being the active principal, for the past fifty years, the sequence in which the amino acids are arranged, in what is so commonly referred to as " thyroglobulin," still remains a mystery. The gland bears a certain resemblance to the parathyroid glands, which is doubtless due to the two having originated from the same, the radiant portion. But, nevertheless their actions are opposite, a difference that enables the extracts of each to be used to antidote the hyperactivity of the other. On the other hand the gland bears a relationship to the *Pars intermedia,* which originated from the attractive portion. How thyroxine acquired its diphenyl ether group; and, how the compound developed into nor-adrenalin, should this have been the first compound to have been formed in the attractive portion, still await solution. Once formed, nor-adrenalin becomes methylated by the amino acid methionine, and on this happening the place held by the thyroid gland is taken by the chromaffin system.

## CHROMAFFIN SYSTEM

This system originated from the attractive portion. It is composed of the medulla of the suprarenal glands in the retro-peritoneal glands, the organs of Zuckerkandl, the carotid body and the posterior lobe of the pituitary gland. All the extracts have an expansive action, or exhibit the function of attraction. The only known gland to possess polypeptides, is the posterior lobe of the pituitary gland; and, this I attribute to the gland having originated from the attractive end of the third junction. The polypeptides are oxytocin and vasopressin both of which contain the disulphide group, that unites in them certain of the amino acids as it does the two polypeptide chains in insulin.

The active group in what are most commonly referred to as adrenergic compounds, is the amino-methyl group. The nitrogen is the most active part of this group, and the part that is responsible for the exhibition of attraction by the group. This exhibition of attraction by the group is more marked in nor-adrenalin than in adrenalin, as is known by the vascular constrictor action of the former exceeding that of the latter. Adrenalin has in part a vasodilator action, which it owes to the donation of the methyl group to the compound by methionine. This damping down of the expansive action of nor-adrenalin by the addition of the contractile acting methyl group, was doubtless effected by Nature. Not only for the purposes of balancing the exhibition of the two reactive functions of radiation and attraction, but also for that of preparing for the attracted " activity " becoming stored in the product. A storage that became necessary if the product were to progress, or evolve.

## ANTERIOR LOBE OF THE PITUITARY GLAND AND PARS INTERMEDIA

These glands originated from the attractive end of the junction between the attractive and store portions. The glands exhibit the functions of attraction and storing, and, in addition, they are connected with the expansion the store portion undergoes, in the description of the pulsatory cycles. This three-fold action of the anterior lobe of the pituitary gland, is the cause of the organs division into the basophilic, chromophobic and acidophilic parts. Several polypeptides have been isolated from the glands, and the sequence in which the amino acids are arranged has been discovered in some of them, particularly in those which are connected with the expansion the store portion undergoes. Through the connexion the basophilic part has with the chromaffin system, both the anterior lobe of the pituitary gland and the *Pars intermedia* are brought into relationship with the thyroid gland, particularly with its attractive end, because the first effect extracts of this part have is to lower the iodine content of the gland. Presumably the area in the third junction, from which the chromophobic part originated, is where " activity," after having been attracted, is stored. Should an amino methyl be found to be the group responsible for the attraction, it is tempting to suggest that the storing may be the result of the trivalent nitrogen atom becoming pentavalent, and having further methyl groups attached to it by methiomine. In the light of the influence the acidophilic part of the anterior lobe of the pituitary gland and the *Pars intermedia* have on the expansion the store portion undergoes, in the description of the pulsatory cycles; and, in which the nitrogen atom plays no part; the possibilities arise of the methyl being the only active group to be retained in this part; and, of an addition being made of an hydroxyl group. Because both of these groups are to be found in the 13- and 17-positions respectively of the steroids, which constitute the basis of the store end of the third junction. When it is remembered that expansion of the store portion is the first physico-chemical change the protein undergoes, in the description of the pulsatory cycles, and that owing to the variability of " climate " no limit can be placed on the grades of expansion the portion may be made to undergo. A new look needs to be given to the existence of a relationship between the several hormones, of which the glands are thought to be the possessors, and the grades of expansion the portion is capable of being made to undergo. Particularly when the action of most of the hormones may be antidoted by extracts of the glands next to be mentioned.

## PINEAL AND PARIETAL GLANDS

These glands originated from the attractive end of the third junction. They are the representatives, in the vertebrates, of the right and left median eyes, in the invertebrates. In the segmented worms these median eyes became replaced by lateral eyes, a replacement which resulted in the original eyes retrogressing. To this retrogression in which the glands, into which the eyes evolved, became involved, I attributed the contractile action the extracts of the glands possess. This contractile action affects both the attractive and store portions, and as in the case of those obtained from the anterior lobe of the pituitary gland and *Pars intermedia*, it is more pronounced on the store than on the attractive portion. The extracts contain polypeptides, but up to date there are none in which the sequence in which the amino acids are arranged has been discovered. In recent years a chemical compound, existing in large quantities, has been found in the pineal gland, to which the name " melatonin " has been given. This compound is the contractile-acting form of the expansive-acting serotonin, which is a common constituent of the brain.

In the early brain the pineal and parietal organs are connected with the roof, and the anterior lobe of the pituitary gland and the *Pars intermedia* with the floor. What has now to be ascertained, is whether the serotonin content of these two glands is the equal or not of that of the pineal gland of melatonin. Because, it would go a long way to establish the close relationship, which exists undoubtedly between these two pairs of ductless glands.

Another relationship demanding clarification, is the one existing between the questionable contractile action of the chromophobic part of the anterior lobe of the pituitary gland and that of the pineal gland.

When I found that the extracts of the anterior lobe of the pituitary and pineal glands have opposite actions on the attractive and store portions, and counteract each other, I assumed that the original action of the sense-organism was an expansive one. And that the part of it acquiring a contractile action, to produce the contractile-acting ductless glands and the parasympathetic part of the autonomic nervous system, arose from the replacement of the median by the lateral eyes, and the regression the former underwent as a consequence. I failed to take into consideration the contraction that automatically follows expansion, and which, in this instance, might have been caused, or at least assisted, by the contractile action of the chromophobic part of the anterior lobe of the pituitary gland. There can be little doubt, that the regression the median eyes and their nerves began to undergo in the segmented worms, had an influence on the development of the contractile part of the sense-organ mechanism, and the ductless glands and nervous systems into which it evolved. But nevertheless, knowledge is needed on the roles automatic contraction, and the contractile action of what evolved into the chromophobic part of the anterior lobe of the pituitary gland, had on the division of the sense-organ into its expansive and contractile parts. From the way in which the steroids in the store portion developed (*vide* p. 25), it is possible that acetylcholine came to be the most important compound in the complex constituting the basis of the chromophobic part of the anterior lobe of the pituitary gland, and through this of the parasympathetic part of the autonomic nervous system.

# EPILOGUE

ENERGY, or "activity," is the matter responsible for the generation of every product in the universe, and for the way in which every product behaves. The generation resulted in every product, with the exception of two, becoming composed of three portions, the store, the radiant and the attractive, which appeared in this order; and of which the first is unreactive compared with the other two, both of which are reactive. The two exceptions are the hydrogen atom and the larval form of the hollow-gutted animals, which are held to lack the store portion, in the form of the neutron in the former and the mesoderm in the latter.

The way in which every product behaves is controlled by energy, or "activity," describing cycles in the product, in the process of which the product either progresses or regresses. The progress and regress arise from the variability of energy, or "activity," causing differences to occur in the physico-chemical changes which the products are made to undergo in the description of the cycles. The products are divisible into atomic, crystalline, or chemical, colloid and vegetable and animal; and the cycles into pulsatory and secondary. In the pulsatory cycles the store and attractive portions expand and contract, whilst the radiant portion contracts and expands. Progress is registered when more "activity" is attracted by the attractive portion when it expands, and the store portion when it contracts, than is liberated and radiated from, and therefore lost to the product, when the store and radiant portions expand.

Ninety-two products are known to have naturally been generated in the atomic section, but it is not impossible that this number had been exceeded in the past, as a further ten have artificially been produced in the laboratory. The representatives of the store, radiant and attractive portions in atoms, or elements, are the neutrons, protons and electrons, to which must be added the other subatomic particles. As the ladder of the elements is scaled, the number of the neutrons grows progressively to exceed that of the protons and electrons, the numbers of both of which keep in step. From the progressive excessive increase in the number of the neutrons, the inferences may be drawn, that, in the description of the pulsatory cycles in the atoms, the store portion grows progressively more over-contracted and the radiant portion over-expanded. Owing to the first of these physico-chemical changes being unreactive, and only the second being reactive, account is given for seventy-six of the ninety-two atoms exhibiting predominantly the function of radiation, and being metals. And the last six to be generated, including all those which are artificially produced, being radio-active, in which state helium atoms, electrons and gamma rays are radiated from the radiant portion. Ten products were generated which exhibit predominantly the function of attraction, and are non-metals; and, six which exhibit predominantly the function of storing, and are inert gases. Of these six, in only one, helium, the second atomic product to be generated, are the neutrons, protons and electrons evenly balanced. The product contains two of each. The respects in which more knowledge is required of the elements, lie in the failure of the hydrogen atom to possess a store portion, if the theory be correct that it was from this portion that the radiant portion developed and from the latter the attractive portion. And in the nature of the factor, or factors, which caused the area in the atoms to undergo the principal physico-chemical changes in the description of the pulsatory cycles in the products, to be the junction between the store and radiant portions.

In the light of finding fungi to be the regressive products of the *Algae*, and bacilli, sub-bacilli and cocci of the fungi; may not another look show that the sub-atomic products are regressive forms of the elements, rather than the bricks which were used in their manufacture. Should the latter view be found to be the correct one, an explanation would be afforded for the number of subatomic particles, which are expected and awaited to be discovered in the future. And it would fit into the limitless grades of expansion and contraction the portions of the products are caused to undergo, in the pulsatory cycles which " activity " describes in them.

The area in the crystalline, or chemical, products, the most affected by the principal physico-chemical changes which the description of the pulsatory cycles in the products makes them undergo, is the junction between the attractive and store portions. Because, in this second section into which the products of the condensation of " activity " are divisible, " activity's " aim would appear to be to neutralise the reactivity of the radiant and attractive portions, by exhibiting as predominantly, as lies in its power, the unreactive function of storing. Three attempts were made to achieve this end. The first, by metallic compounds combining with non-metallic compounds. The second, by dissimilar non-metallic atoms combining to form chemical compounds. And the third, by carbon in the formation of chemical compounds in which the atom became the nucleus.

The affinity metallic compounds show for combining with non-metallic compounds, to reach a more unreactive physico-chemical state, needs no explanation as the exhibitions of radiation and attraction become automatically followed by the one of storing. The exhibition of storing follows immediately the one of attraction, and herein lies the explanation for the facility with which dissimilar non-metallic atoms are able to unite to form chemical compounds. Carbon owes its entry into the competition, if the matter can be considered to be of this nature, first, to its being the atom best equipped to exhibit all three functions with almost equal facility. And second, in the order of storing, radiation and attraction. Although the carbon compounds came nearest to reaching the goal, none approached so close as hydrogen had been able to do in the first, or atomic section, on the occasion progress led to its evolving into helium. In no other product did its three portions ever become so well balanced as they did in the second atom to be generated, helium. The facts that helium is the form in which hydrogen is radiated from the first reactive portion in the radio-active metals. That the radiant portion has to over-contract before it can over-expand. And that the half of a pulsatory cycle, in which the radiant portion expands, needed to be described, in addition to a complete cycle, before the third element, lithium, was able to evolve, have occasioned me to wonder if the universe, which is now supposed to have had no beginning, did not originally exist in a state even more stable than the one in which helium appears to flourish. Or in short, that if it is not inherent in "activity," to exhibit predominantly the unreactive function of storing, in preference to either of the two reactive functions of radiation and attraction. If the case is as surmised, it was destined not to materialise. Because, not only did the formation of chemical compounds fail to produce the results expected, but also the combined three portions of the progressing products were enabled to reach an independence,only one at a time had previously been able to do. This independence was gained only by those chemical compounds, which were evolving in the water, and it was effected by the medium becoming electrolysed into an hydroxyl group and a hydrogen atom. This group and atom formed an inner and outer ionic layer enveloping the product. The envelopment of the product by these two electrical layers, not only gave the product an independence it had never previously enjoyed, but also the layers acted as protectors, by being the part of the product, in which its actions with other products are engaged. The formation of these two layers resulted in the crystalline, or the chemical, products progressing into the colloids; and the principal area in the products shifting once again to the junction between the store and radiant portions. This shift arose from the liberation of the hydrogen in the expansion the store portion undergoes, in the description of the pulsatory cycles, and its acceptance by the radiant portion as it contracts. In the second half of the pulsatory cycle the hydrogen atom is donated by the radiant portion when it expands, to the store portion where it reactivates the hydroxyl group, which became unreactive when the portion contracted. This order of events was the factor which made the hydroxyl group the inner layer, and the product so-called "negatively charged." Possibly the first crystalline compound to be rendered colloid was silicic acid. The reasons for selecting this compound are. First, that the sandstone rocks were the first to disintegrate. Second, that the compound is to be found in abundance in the cell-wall of some of the earliest vegetable and animal products to be generated, the diatoms and radiolarians. Third, that freshly prepared colloid silicic acid may take the place of the amboeceptor in the most important of all the immunity tests, the complement-fixation test. Indeed, this third reason suggests, that at the time the colloid products were evolving into the vegetable and animal products, a competition existed between silicon and carbon, for which of the two atoms was to hold the highest position of any in the generation of the animate products. Should such a competition ever have existed must be open to question, but if one had done, no doubt can arise on the matter of carbon having been the victor. Silicon lost owing to the colloid product becoming hydrated in time, in the process of which the product is made to regress into the crystalline state. This is the state in which the product exists in the diatoms and radio-larians and other animate products. The acceptance of the liberated hydrogen, and its donation by the radiant portion in the two halves of the pulsatory cycles, gave this portion the lead of the other two portions which continued to increase as further progress was recorded. But nevertheless, which never ceased to be influenced by the grade of expansion and contraction the store portion is made to undergo, before the radiant portion contracts and expands. This lead which began with the exchange of hydrogen between the store and radiant portions, extended to that of the same atom between the attractive and radiant portions. Then followed the exchange of active groups and chemical compounds between the two pairs of portions. And finally, the invasion of organs, which had originated from the radiant portion, into those which had originated from the other two portions. The progress which was made in the lead the radiant portion gained of the other two portions, coupled with the protection which was afforded by the envelopment of the three portions by the two ionic layers, gave the products a chance to evolve it had never previously been offered. This product, in its pristine form, is the one to which the name " protein " has been given. But it is the product which may arbitrarily be fixed as having been the last to have been generated in the colloid section.

The time when the three portions became sufficiently equilibrated, to permit of organs originating from them, also needs arbitrarily to be fixed. The period fixed is the one in which the colloid products evolved into the vegetable and animal products. In this period the protein existed in the form of cytoplasm, and the product in that of a cell divisible into cytoplasm and a cell wall, and containing an organ, the nucleus, which had originated from the radiant portion. At this stage of evolution the product was animate, but not yet divisible into vegetable and animal. The progress made by a colloid into an animate product, was effected in water, and, it was not until much later that the animate product became divisible into vegetable and animal. The more thought that is given to the evolution of the animate products, the clearer it seems to become that they evolved as such, and only became divisible into vegetable and animal as a result of having been forced to live under different conditions. And the closer these conditions are examined, the more they are found to be linked with the relationship which came to be established between the three portions of the products. The depth in the water in which the animate products were generated. Whether they became fixed to rocks, where autonomous arrangements had to be made for the adequate description of the secondary cycles. Or were free to move about, to breath air and to live on food which had already been prepared for them. Whether they came to live on land, where drying conditions may prevail, had a tremendous influence on the ways in which the animate products evolved. In spite of the differences between these conditions, they had the effect of dividing the products into classes; and, what affected the division were the physico-chemical changes the differences made the store portion of the products undergo. The greater the depth, the more fixed the position and the drier the condition, the more over-contracted the store portion tended to become. Whereas when the opposite conditions prevailed a more equilibrial relationship between the three portions was enabled to be maintained. Over-contraction of the store portion causes the radiant portion to over-expand and the attractive portion to over-contract, in the processes of which the organs which originated from the portions vanish by liquidation and absorption respectively. So long as the three portions are maintained in equilibrium, the store portion is enabled to exhibit predominantly the function of storing, without seriously affecting the other two portions; and, of effecting a closer relationship with one of the reactive portions on one occasion and with the other on another, without damaging the product. In short, those products which were made to retrogress by the store portion over-contracting, were those which became the vegetable products, whereas those which were enabled to progress, by the store portion being allowed a freedom of action, were those which became the animal products.

In the change the unicellular animate products underwent into the multicellular products, the cytoplasm, enclosed within its cell wall, escaped to become dispersed into the colloid protein particles in the sap and blood. And to the physico-chemical changes the store portion of these particles was made to undergo, by the description of the pulsatory cycles in them, may be attributed the fundamental characteristics of both the vegetable and the animal products. These characteristics include the radial symmetry of the latter. Possibly the most successful attempt which was made, in the animal products, by the store portion to share equally its favours between the radiant and the attractive portions, was the origin, from the very centre of the portion, of the notochord. Nevertheless, despite the origin of this organ having been responsible for the division of the animal products into one of its largest groups, the *Chordata,* it failed to prevent the unequal distribution of the favours in the reptiles. This unequal distribution resulted in the store portion forming a closer association with the radiant than with the attractive portion in the birds, and the reverse in the mammals.

Attention is called to this example of the difficulty of reaching equality in Nature, to conform the variability and unpredictability of " activity," in the form of " climate," which is the factor responsible for the way in which the store portion behaves in the description of the pulsatory cycles in protein. Following the origin of organs from the protein in the animate products, the necessity arose for the establishment of a mechanism able to keep the two equilibrated. The mechanism which became established, was the development of the secondary from the pulsatory cycles. And it was this development which resulted in the progress and regress in the inanimate products, becoming replaced by health and disease in the animate products. For health to be maintained the secondary cycles need to be normally described, but as the way in which they are described is determined by that in which the pulsatory cycles have previously been described. And as the quality and quantity of " climate " are the factors which determine the way in which the pulsatory cycles are described, difficulties will never cease to be experienced in distinguishing between health and disease; and in judging the efficacy of the attempts which are made to prevent the former developing into the latter.

Health is best maintained by insisting on the air breathed and the food eaten being of the highest quality, and on the conditions under which life is lived being kept as natural as possible. The future may show that the immunological methods now being employed to prevent health being changed into disease, should be reserved for use only in those who are potential victims of the manifestations the preparations are hoped to prevent.

In deciding whether an animate product is healthy or diseased, determining the physico-chemical state, in which the protein exists, should be looked upon as being one of the most important of the examinations to be made. Because, if undertaken as a routine, the examinations show how " climate " is acting; and give a measure of the size of the area over which it is acting. Charting the results obtained, give the world the pictures it needs of the forms of the types of mosaic in plants, distemper in animals and influenza in man, which are the acute, or seasonal, manifestations occurring in the animate products; and which start them on the downward path from health to disease. The factor operating to continue the decline, is the acquisition of pathogenicity by the fungi which regressed from the *Algae*, and the bacilli from the fungi, the latter of which, in addition to remaining in the soil, have become resident in the intestinal tract of animals and man. It is to this pathogenicity, that one has to turn to find the principal cause of the subacute manifestations of disease. This further establishment of the state of disease results, either in the store portion remaining over-contracted or becoming over-expanded. To cause, in the first case, atrophic lesions and, in the second case, hypertrophic lesions to arise in the organs, which originated from the portion, in the chronic stage of disease.

In addition to manifestations and stages, disease is further divisible into morbid conditions of which there are two, inflammation and cancer. This third division arose from the junction between the store and radiant portions being the area in the protein in which reside most of the assistants of the leaders in the metabolism the products undergo. And from which originated not only the first organs, but also those which are involved in the metabolism of the whole product, and not solely with that of the portion and the organs which originated from it, to which they belong. In short, this junction may be regarded as the headquarters of the protein metabolism. The grade of over-contraction the junction undergoes in the acute stage of disease, determines the organ in which the manifestation is to appear; and whether the so-called " virus " becomes formed in the area in the cytoplasm of the epithelial cells, or in the cytoplasm and nucleus of the nerve cells. In the subacute stage the grade of over-expansion the junction undergoes, decides the organ, other than those which originated from the store portion, in which the manifestation is additionally made to appear. If the protein is to clot, and whether the mononuclear leucocytes are to replace the polymorphonuclear leucocytes to form the second line of defence. Should the expansion be reflected beyond the protein on to an organ, the latter is the nucleus, which is stimulated to form a new growth. The new growth makes its appearance most commonly in the skin, on account of this being the organ which is most likely to be attacked when the grade of over-expansion the junction undergoes comes to be reflected on to the nucleus. When the over-contraction the junction undergoes in the chronic stage reaches the grade, where it comes to be reflected on to an organ, the one attacked is the nucleolus of the nucleus. Cancer is the morbid condition produced;

and the larger organ, found to be the bearer of the damaged nucleus, is governed by the extent to which the junction has been made to over-contract.

In the description of the sexual reproductive cycle, should the junction be more contracted than expanded the greater is the likelihood of the progeny being male; and in the case of the reverse, female.

In the description of the therapeutic cycle, the aim is to restore to its normal physico-chemical state the protein which has been made by the causative agents of disease to undergo abnormal physico-chemical changes. In the acute stage the goal is more easily reached with the assistants, which reside in the junction between the store and radiant portions, and which act by expanding the over-contracted area. These assistants in the acute stage are potentially dangerous, and the more so the closer their chemical configuration approximates that of the compound the unicellular regressive forms of the *Algae* stand most in need; and, have the greatest difficulty in preparing for themselves. The danger lies in rendering immune, and increasing the parasitism of the fungi, bacilli, sub-bacilli and cocci, which become pathogenic. In the subacute stage aim should be made at expanding the over-contracted store portion, and contracting the over-expanded radiant portion. The result can best be obtained by combining the steroid preparations with either the assistants or their leaders, which have their home in the radiant portion. In addition, valuable use can be made of the extracts obtained from the micro-organisms, which play a role in the production of the abnormal physico-chemical changes the protein is made to undergo in this stage.

It is in the chronic stage of inflammation that the greatest difficulty is most likely to be experienced, in selecting the most efficient remedy. First, because knowledge needs to be had of the portion of the protein which is being made to undergo the greatest abnormal physico-chemical change; and second, whether this abnormal physico-chemical change be one of over-expansion or over-contraction.

At present there are available no remedies, which are known to have a contractile effect on the over-expansion the store portion tends to undergo, as age advances. In view of the slight retarding effect, extracts of the pineal gland (acervulin) have upon growth, this might prove to be the best quarter, in which to search for such a preparation. When the portion is over-contracted more reliance is placed upon the expansive action of the glycosides.

The remedies best able to contract the chronically over-expanded radiant portion, are the active principles of the para-thyroid glands, the thymus and the islet cells of Langerhans. When the portion is over-contracted, the preparations found to have the most efficient expansive action, are the internal secretions thyroid gland and the chromaffin system.

From the little work I have done on the pineal gland, I feel confident that once all its active principles have been isolated, one or more will be found that are better able to contract the over-expanded attractive portion, than the oral preparations at present in use, and of which the most suitable are the hypotensive remedies used singly, or in combination with one of the chemical compounds belonging to the thiazide series. All these oral preparations have the potential danger of over-contracting the attractive portion, which today appears to be the worst abnormal physico-chemical change any of the three portions may be made to undergo. This statement is made, judging from the galaxy of drugs which have recently been introduced to expand the over-contracted attractive portion, and which can be added in mentioning them, that all are vested with the potentiality of aggravating the over-contraction. These drugs include the hypnotics, the tranquilizers, the antidepressants and anticonvulsants, and those which are employed in the treatment of Parkinsonism and the production of anaesthesia. The questions arise—why has over-contraction of the attractive portion become the outstanding abnormal physico-chemical change the protein in man has been made to undergo in recent years; and psychiatric, the main manifestations to which it has given rise? The most probable answer is that the expansion the store portion has been made to undergo in the description of the pulsatory cycles in the protein during this period, has been of the grade which causes the area in the attractive portion, from which the central nervous system originated, to over-contract. In support of this answer being the correct one, has been the replacement, during the period, of the forms of the cutaneous type of influenza by the forms of the nervous type. Should this answer be found to be correct, and that a relationship does exist between the nervous type of influenza and the psychiatric manifestations, no clearer demonstration can be given of the necessity to chart the periodic physico-chemical changes the protein undergoes. And to look at disease as a whole instead of concentrating attention on each of its several parts.

Temporary success in the treatment of cancer has been achieved by expanding the over-contracted store portion, with the cortico-steroids. And with expanding and contracting the over-contracted and over-expanded radiant portion, with modified constituents of the area in the portion from which the nucleus originated. But it is doubtful if success, should it ever be obtained, will be reached until the polypeptide responsible for the generation of the nucleus, especially of its nucleolus, has been discovered and isolated. In the cases of cancer in which the primary growth has been removed or treated, secondary growths have been prevented from arising, following the parenteral use of acervulin, which acts by contracting the over-expanded store portion of the protein.

# INDEX

# INDEX